ARVO PÄRT

Sounding the Sacred

Arvo Pärt

SOUNDING THE SACRED

Peter C. Bouteneff,
Jeffers Engelhardt,
and Robert Saler
EDITORS

FORDHAM UNIVERSITY PRESS   NEW YORK   2021

Library of Congress Cataloging-in-Publication Data available online at https://catalog.loc.gov.

Printed in the United States of America

23 22 21   5 4 3 2 1

First edition

Contents

I. INTRODUCTION

1. Arvo Pärt and the Art of Embodiment 3
   Peter C. Bouteneff, Jeffers Engelhardt, and Robert Saler

2. The Sound — and Hearing — of Arvo Pärt 8
   Peter C. Bouteneff

II. HISTORY AND CONTEXT

3. Sounding Structure, Structured Sound 25
   Toomas Siitan

4. Colorful Dreams: Exploring Pärt's Soviet Film Music 36
   Christopher J. May

5. Arvo Pärt's Tintinnabuli and the 1970s Soviet Underground 68
   Kevin C. Karnes

III. PERFORMANCE

6. The Pärt Sound 89
   Paul Hillier, in conversation with Peter Bouteneff

7. The Rest Is Silence 107
   Andrew Shenton

IV. MATERIALITY AND PHENOMENOLOGY

8. Vibrating, and Silent: Listening to the Material Acoustics
   of Tintinnabulation   129
   Jeffers Engelhardt

9. Medieval Pärt   154
   Andrew Albin

10. The Piano and the Performing Body in the Music of Arvo Pärt:
    Phenomenological Perspectives   177
    Maria Cizmic and Adriana Helbig

V. THEOLOGY

11. Presence, Absence, and the Ambiguities of Ambiance:
    Theological Discourse and the Move to Sound
    in Pärt Studies   197
    Robert Saler

12. The Materiality of Sound and the Theology of the Incarnation
    in the Music of Arvo Pärt   208
    Ivan Moody

13. Christian Liturgical Chant and the Musical Reorientation
    of Arvo Pärt   220
    Alexander Lingas

14. In the Beginning There Was Sound: Hearing, Tintinnabuli,
    and Musical Meaning in Sufism   232
    Sevin Huriye Yaraman

LIST OF CONTRIBUTORS   243

INDEX OF TERMS   247

INDEX OF PERSONS   252

WORKS BY OTHER COMPOSERS   256

WORKS BY ARVO PÄRT   257

ARVO PÄRT

Sounding the Sacred

I

Introduction

1

Arvo Pärt and the Art of Embodiment

Peter C. Bouteneff, Jeffers Engelhardt, and Robert Saler

Music is sound. For many readers this may seem like an unremarkable statement — a tautology. Yet to a growing number of scholars the understanding of music as sound is a call to examine it from a series of fresh perspectives. Music — the human shaping and perception of sung, struck, bowed, or electronically generated tones, as musique concrete or even as the supposed "silence" of John Cage's $4'33''$ — consists in sound, in an interplay with (relative or metaphorical) silence. Music is a phenomenon of vibrating matter or oscillating energy, materially mediated to fleshly bodies. Only subsequently can it be "heard" in the mind, transduced into an electrical signal from sound waves: That silent audition — music's mental embodiment — remains contingent on the physicality of sound.

There are expansive implications of the basic fact that music is a subset of a wider world of sound, which is one of the primary concerns in the field of sound studies. In contrast to discussions in music studies that limit themselves to matters of composition — notes on pages — rather than the actual material conditions of sound and hearing, sound studies is concerned with the phenomenology and materiality of the auditory event that is sound. Sound studies directs attention to sound in all its dimensions and with far-reaching outcomes (philosophical, cultural, historical, ethical, affective), and, while it strenuously avoids privileging music from among the world of sound, it does invite fresh investigation into the production and experience of music.

The music of the Estonian composer Arvo Pärt (b. 1935) has inspired a growing number of scholarly publications. These are situated primarily in the fields of musicology (analyzing his signature "tintinnabuli" method), cultural and media studies (as the world's most-performed living composer, his audience is uncannily broad within the contemporary classical world), and, more recently, in terms of theology/spirituality (Pärt is primarily a composer of sacred music). For the most part, these books and essays have circled around the representational sonorousness of Pärt's music, writing and listening past the fact that its storied effects and affects are carried first and foremost as vibrations through air, impressing themselves on the human body.

One reason this is of particular significance for Pärt is that, while "Pärt studies" has been reticent on sonic phenomena, it has had a great deal to say about *silence*. Pärt's music, though diverse in its sonic effects, is commonly experienced as quiet, stilling, somber, meditative. Some music scholars have flirted with placing Pärt in the category of "minimalists," though not without impassioned rebuke from others. The music does indeed breathe silence, either by virtue of consciously summoning it through spaces between notes, espousing a reductive, essential compositional style, or evoking silence in more abstract ways.

So Pärt-and-silence is a subject eminently worthy of attention. But what of Pärt-and-sound?

That essential question was at the heart of the conference "Sounding the Sacred," organized by the Arvo Pärt Project, at St. Vladimir's Seminary, held May 1–4, 2017, in New York. Within the broader task of exploring the ways that spiritual reality is embodied in the material—an apt theme for a Christian faith centered on incarnation and sacramental embodiment—the idea of "Sounding the Sacred" was all but inevitable: How is "the sacred" embodied in the materiality of sound and made to resonate within the body of the listener? And how does this take place through the oeuvre of a composer who is so widely experienced as "spiritual"?

Speakers at the conference represented diverse disciplines, reflecting the collage of concerns within sound studies: architecture, history, acoustics, music studies, cultural studies, psychology, performance and materiality studies, and theology (mostly Christian, though with a significant contribution from Sufi Islam). A selection of these, as well as a few other scholars whose work touches on this conversation in vital ways, are now assembled in this volume in order to carry on the conversation in a sustained and critical way.

As some of its most significant fruits, this interdisciplinary volume promises new insights into the origins not only of the "Pärt phenomenon" but of the elements underlying his post-1976 compositions. Specifically, as the essays by Karnes, Siitan, May, and Albin show, the breakthroughs that Pärt made in the 1970s through tintinnabuli did not come out of nowhere. Even as music scholars have traced some of tintinnabuli's roots to plainchant and early music, there is no question that much of the initial portrayal and "packaging" of

Pärt both in Europe and in the United States has traded on an image of him as not only a sort of musical mystic but also one whose compositional methods and their sonic purity were reflections of a higher sacred order beamed from heaven on a winter's day in 1976. The story is sometimes too simplistically told of how Pärt retreated into "silence" for eight years, writing only film music, which meant little or nothing within his oeuvre, only to emerge suddenly with a tintinnabuli that had no traceable precedent. Several essays in this book demonstrate that tracing Pärt's own relationship with sound in his formative compositional years produces a much more plausible — and indeed far more interesting — picture. The nature of Pärt's eventual break from the strictures of musical modernism and the Soviet experimental scene was part of a broader shift within the scene itself, a shift that replaced orthodoxies about what counted as "true" art with a more curious spirit toward the resources of the past, including spiritual resources and, as Albin shows us, "medieval" relationships with words, texts, bodies, sounds, and silences. This book stands, among other things, as an argument that Pärt's genius lies not in an ex nihilo innovation but rather in the creative adaptation and evolution of nascent themes that shaped an entire generation of musicians in his context — something that by no means undermines the uniqueness of his oeuvre and its impact.

This is an important point: Just as sound studies seeks to "de-Platonize" music by considering it as an embodied phenomenon and not just abstract notes in the ether, so too does contextualizing Pärt as a historically embedded artist working with and adapting existing themes and techniques help to debunk a false narrative about the nature of his achievement in order to allow unforeseen contours to emerge. Contingency and embodied finitude, not eternal ether, is the stuff of music and its creation. Likewise, it is worth demystifying — as Bouteneff does — the notion of a single "Pärt sound" by highlighting the complexity and diversity of Pärt's tintinnabuli works. Put simply, an ahistorical or monochromatic Pärt is both implausible and ultimately less impressive than a contextually sensitive artist who is of his time in such a way that his work touches our era with such perception and pathos.

As the contributions by Shenton, Hillier, Cizmic and Helbig, and Engelhardt illustrate, though, thinking about the embodied character of the Pärt sound should focus on more than just the composer. Phenomenological inquiry into the performance of Pärt's music, the acoustics of tintinnabuli as it materializes in sound, and the concrete techniques and bodily positions needed to achieve excellence in performance is also necessary. These chapters seek to move past the notes on the page and consider the bodies — human and object — that are arrayed to produce the distinctive sonic characteristics of a "Pärt piece." While these discussions, like the others, must necessarily be wide ranging, the cumulative effect is one of rigorous focus on Pärt's sound as a fully embodied reality, with all the fragility, contingency, limits, and possibility therein.

Any complex conversation about the Pärt phenomenon needs to be interdisciplinary. It must reach and be reached by cultural studies, history, music studies, media studies, and other disciplines. Fortunately, sound studies by its very nature is already interdisciplinary: To discuss bodies is to discuss the epistemological, political, and phenomenological worlds in which bodies move, create meaning, and experience vibrations. Sound studies does not prescribe a single methodological starting point, and thus neither do the essays in this volume.

Within that interdisciplinarity, once a composer of sacred music is involved, religious studies must be a part of the conversation. But what about theology, specifically? Theology occupies a contested role within the field of religious studies more broadly. Given the prevalence of the "mystifying" narratives surrounding Pärt, there might be reason to be more than usually suspicious of theology's role in a conversation about embodied sound. Here some more specific commentary is in order.

The study and reception of Pärt's music through the lens of Christian theology has never been without controversy. There is a wide middle ground that considers theology an innocuous incursion into the reception of Pärt's work; this is flanked by two more extreme camps. One of those camps sees theology as a violent ecclesiastical colonization of an otherwise pure and spiritual experience (the less said about which, the better). The other camp sees it as utterly inevitable, given the overtly sacred content of Pärt's oeuvre and its underlying texts. In any case, anyone venturing to bring Pärt's music and Christian theology into conversation does best to stick to making propositional statements, not dictatorial pronouncements. In that view, to posit a causal relationship between theology and Pärt would be risky and impossible to prove, while suggesting correlative connections can awaken potentially new insights without binding either the reader or the composer to a fixed interpretation. In this way, one is most likely to maximize potential insight and minimize the programming or regulation of any listener's spiritual experience. That propositional stance has undergirded the efforts of the Arvo Pärt Project, whose stated function is to serve as a forum for exploring the connections between the music, the composer's faith, and the sacred texts to which the vast majority of his music is set.

These relationships raise the methodological questions — by now well worn but still significant — surrounding the relationship between art and artist. Is that relationship consequential? Does insight into the artist shed any light on the artwork? These questions place another imperative on anyone tempted to "own" Pärt's music as a result of knowing Pärt and his faith world: an imperative to present any findings or opinions as offerings that can be accepted or not. Yet, taking on board this encouragement to walk gently, it would be disingenuous to ignore Pärt's identity as someone who composes sacred music and who does so not by way of abstraction or homage but from a location of fervent belief. The fact that this location is set within a pluralistic, secular world means our task to seek to describe it critically in a volume such as this one is all the more important. Essays by Bouteneff, Shenton, Saler, Yaraman, and Moody each in their own way make the case that theology is a generative conversation partner in this discussion.

Theological considerations likewise play a role in the concluding essays of this volume. They explore some of the diverse spaces within which Pärt's music can be said to resonate, whether sonically, thematically, or affectively. Lingas studies Christian liturgical chant—which is not a direct source for Pärt's compositions, as is frequently and erroneously supposed—as a practice that shares an ethos and textual sensibility with Pärt's music. Through a creative engagement with Sufi poetry, Yaraman discovers resonances both in the "bright sadness" characteristic of Pärt's compositions as well as in the confounding equation ($1+1=1$) that has been used to describe the sonic effects of tintinnabuli.

The present volume assembles a series of forays into the sound-focused study of Pärt's music. While no such collection could claim anything like finality or comprehensiveness, these essays demonstrate—at a minimum—that this critical and methodological orientation is capable of yielding many new insights, not only into Pärt's work and impact but also into the ways in which "the sacred" is potentially conveyed and received through sound.

The editors wish to thank the following for their support of the conference that gave rise to this volume: St. Vladimir's Orthodox Theological Seminary (home to the Arvo Pärt Project and the Institute of Sacred Arts), Fordham University's Orthodox Christian Studies Center, and the Henry Luce Foundation. Nicholas Reeves and Lisa Radakovich Holsberg were our colleagues on the organizing committee and played an instrumental role in shaping the conference. In addition we wish to thank the Arvo Pärt Centre for permission to reproduce archival material and for their overall support, Universal Edition, Eres Edition, Joonas Sildre, the Office of the Provost and Dean of Faculty at Amherst College, and the faculty, staff, and administration of Christian Theological Seminary. Thanks also to the editing and production team at Fordham University Press.

2

The Sound—and Hearing—of Arvo Pärt

Peter C. Bouteneff

The study of sound concerns the relationship of two distinct yet interdependent things. One is the physical phenomenon of the movement of air; the other is the hearing subject. In order for air in motion to be called "sound," it requires a hearer.[1] The reception, in the physical body of a sensate subject, is as much a precondition of sound as is the "sound wave" that is received. That is the definition I am working from, at least, and it means that if a tree falls in the forest and there is nobody there to hear it, it actually does *not* make a sound: One presumes that waves are generated in the surrounding atmosphere by the plummeting tree and its impact, but that is all.

It follows that, in considering the vast, varied, and affecting repertoire of Arvo Pärt, the turn in attention toward its actual sound would involve addressing fundamental questions within roughly these same categories, of sound and hearing subject. The questions may therefore emerge as follows: First, what is the sound of Arvo Pärt?[2] And second, what are we hearing when we hear Arvo Pärt?

These two questions are not without overlap. The first is directed outward. It has to do with the production of the music, together with the sonic product itself. Within that inquiry, I am interested in asking how one might identify "the sound of Pärt," beginning by challenging the notion that there actually is one, or only one. The second question is directed inward. Here, even as I retain an interest in the product as such (and even the production process), the questions surround the listener's reception, involving both aesthetic/affective as well as hermeneutical considerations. Especially when it comes to the spiritual character of the music that so many listeners identify, I ask how that finds its way through their corporeal, neural, and spiritual senses.

The intention here is to come away with an enriched sense of this music and how it operates, through the medium of sound, in its receiving subjects — its listeners.

On the Sound of Pärt

Human beings frequently generalize about large and variegated categories, such that we can say common things about vastly diverse constituencies. We might make observations in common — about men, women, Canadians, dancers, artisans, postal workers — even if none of these are remotely monolithic in character. We do so, sometimes to a disarmingly humorous effect, and sometimes with life-threateningly reductive results. No such generalization can go without a process of querying and qualification.

Along these lines, I am frequently struck when people — including sophisticated critics — generalize in this way about Pärt's music. We hear comments along the lines of, "I love Arvo Pärt! His music is so quieting, so stilling. It understands me, it nourishes my soul." Alternately, "I hate Arvo Pärt! It is empty music: minimalism masquerading as profound. What is so significant about descending A-minor scales?" I ask whether it is in fact possible to identify the actual subject of such assessments.

Which Arvo Pärt do you love or hate? Is it *Solfeggio? Orient & Occident? In principio? The Deer's Cry? Dopo la vittoria? Sarah Was Ninety Years Old?* These compositions all create vastly different sonic landscapes, coming down in different places along the apparent binaries of loud/soft, simple/complex, "tonal"/"atonal." Nothing else in Pärt's work sounds remotely like any of these compositions. Granted, it is possible to group several like-sounding works within Pärt's oeuvre, and we will explore this fact here. The pieces that adhere most strictly to the fundamental "tintinnabuli" principles are capable of sounding more like one another, especially once a listener becomes attuned to the melody-triad dynamic in its sparer articulations. Yet I would make the claim that Arvo Pärt's total output, even if we limit ourselves to the post-1976 works, is more diverse, tonally and sonically, than that of virtually any of the classical composers up until (and perhaps even including) the twentieth century.[3]

Part of why people think there is a clear-cut and unified "Arvo Pärt sound" is that film soundtracks, music set to dance, television advertisements, and other multimedia works repeatedly draw from the same five pieces: *Spiegel im Spiegel; Fratres;* "Silentium" from *Tabula rasa; Cantus;* and *Für Alina.* These compositions account for the vast majority of the soundtrack use of Pärt — *Spiegel* chief among them by a long stretch — to the near exclusion

of other works.[4] So that even though *Spiegel,* "Silentium," and *Cantus* are
three quite different sound experiences (despite their common roots in tintinnabuli), they together constitute the gentle, stilling, reductive, sad yet con-
soling "Arvo Pärt sound"—at least for those whose experience with Pärt relies
on his being sounded in conjunction with other media. For those who know
even a little more of the composer's work, they will possibly experience the
multiple sound universes of the above pieces, plus *Magnificat.* If they have
been at the right ambient/drone/New Wave concerts and/or raves, perhaps
they also know *Nunc dimittis.*[5]

In the face of the diversity of the oeuvre, however, the legendary Estonian
conductor Neeme Järvi is still able to argue that there is a Pärt sound, or a Pärt
identity: "You put on a piece and you can tell at once it is Pärt—even the early
pieces," he has said.[6] To make such an observation might require someone
with Neeme Järvi's musical insight and familiarity with Pärt, which describes a
small subset of people. But is Järvi right, or is this just something that one says
in admiration for a composer? Is he operating in a completely intuitive sphere
that eludes all analysis? (Would he really have identified *Orient & Occident*
or *Ein Wallfahrtslied* as Pärt compositions at his first hearing?) In the case of
the sonically varied works of Pärt—limiting ourselves to the roughly one hundred distinct works since 1976—let us see if we might helpfully identify some
factors that do potentially unite his work into a "Pärt sound."

Tintinnabuli

One element that unites most of the post-1976 compositions is of course
tintinnabuli. Tintinnabuli can be understood in quite narrow terms as a
technique—the fundamental rule of the two voices, melody and triad—or in
broader terms, as a sonic world or *topos.* It is both. Or rather, by virtue of the
technique/rule, as well as other features, tintinnabuli becomes that "space"
that Pärt speaks of wandering into, in one of his more oft-cited ruminations.[7]
Leopold Brauneiss, who has analyzed Pärt's use of the tintinnabuli style more
thoroughly than anyone else to date, acknowledges that it is more than the
melody-triad encounter.[8] The characteristics he lists as describing tintinnabuli
begin with "reduction," or limitation to the simplest basic elements of tonality.
("Reduction" is preferable to the fraught term "minimalism.")[9] And in this case
"reduction" means, yes, melody and triad voices. Brauneiss continues, speaking of tintinnabuli as "a new musical *ductus,* mainly in slow tempos, whereby
each note is given individual weight and significance." And finally, he points
out, it is "a highly formalized compositional system in which the melodic and
harmonic progressions are the result of a network of interrelated rules, which

can partly be expressed in formulae."[10] Any account of tintinnabuli has to take seriously the fact that, for all its spiritual freedom, its oceanic character, its simplicity, it is highly formalized. For the most part, rules govern note lengths, note placement, melodic shape, rhythm. Pärt also breaks them, in nearly every composition, but these exceptions only prove the rules.[11]

Tintinnabuli, in all these varied but integrated dimensions, unites nearly all Pärt's post-1976 work. But this by no means allows us to identify a single signature Pärt sound, even if there are some strict tintinnabuli works, such as *Missa syllabica*, *Summa*, *Passio*, *Für Alina*, and most recently *Sequentia*, that might serve as classic examples. The style or *topos* of tintinnabuli is flexible enough, adaptable enough across keys, tonalities, dynamics, levels of complexity, and degrees of adherence to the strict rule, to produce a vast diversity of musical textures: in other words, to produce diverse sound experiences. So tintinnabuli may indeed be cited as a significant uniting factor within a "Pärt sound," so long as we qualify this to embrace a wide diversity.

Logogenesis

Another factor that brings together Pärt's compositions is their basis in text, which has a profoundly formative role in their sound. Only a very small number of Pärt works have no textual roots, most of these being early (and iconic) works dating within five years of the birth of tintinnabuli. This fact on its own would be of limited significance were it not for the fact that Pärt's melodies are, more often than not, shaped by the text's words and phrases. As I will show, the effect of the words on the music owes in part to the rules he establishes for note lengths and emphases based on where and how the syllables fall. Effectively, this means that Pärt compositions will often sound in a way that recalls speech patterns. But their rhythmic patterns and the ways in which they eschew fixed meters (except in cases where the texts themselves are strictly metered, as in *Stabat Mater*, for example) mean that his melodic lines will have an organic sound that roughly mimics human breath and speech. This is significant in suggesting the possibility of a Pärt sound.

I will say more about logogenesis when we approach our second set of questions.

Sound-Consciousness

In identifying the alleged "Pärt sound," it is worth pointing out the composer's demonstrated interest in the world of sound. His work as a sound engineer early in his career undoubtedly enhanced that alertness. There are also numerous quotes one could assemble that testify to an awareness — unique even among composers — of the significance and even centrality of sound in his own life and work. "Sound is my word,"[12] he says, drawing already on an understanding of "word" that recalls the Greek philosophical/theological concept of *logos*, suggesting not a mere word-event but the totality of a principle — especially an expressed principle. Sound, he is saying, constitutes the substance of his self-expression. Stating that idea somewhat more prosaically, he explained his reluctance to say much about his music, or about anything really, by telling a roomful of Catholic clergy seeking out his verbal wisdom, "I apologize but I cannot help you with words. I am a composer and express myself with sounds."[13]

The significance of sound for Pärt is not limited to its instrumentality in his creative and spiritual expression. He realizes that sound has ethical implications. "What is the significance of a sound or a word? The thousands that have flown past our ears have made our receptive apparatus numb. *One should be careful about every sound, word, act.*"[14] He even says, more pointedly, in his widely shared interview with Björk, "You can kill people with sound. And if so, then maybe also there is sound which is the opposite of killing. You can choose."[15]

Such pronouncements testify to the composer's exceptional attentiveness to the sonic world. To these we would add Pärt's relationship with Manfred Eicher, as intense and involved a collaboration between composer and producer as can be imagined, that has testified to the meticulous care taken over the sound of Pärt's definitive recordings since ECM's 1984 release of *Tabula rasa.*[16]

Yet there is a paradox in Pärt's relationship to sound. On the one hand, let us consider his orchestrations, his meticulous choice of instrumentation. They reflect careful, intentional deployment of timbre. On the other hand, we have to consider the number of varieties he has introduced into the instrumentation of specific works over the years. There are no fewer than seventeen different arrangements of *Fratres* listed in his official repertoire. It is among the Pärt works listed as having "no fixed instrumentation"; the others include *Pari intervallo*, *Arbos*, and *Da pacem Domine.*[17]

So along with the painstaking consideration of sound in conceiving his compositions, there is also, in the case of some of them, a considerable freedom. To underscore that point, Pärt seems almost to relativize the consideration of sound and resonance: "Music must exist in and of itself . . . the mystery must be present, independent of any particular instrument." Driving home his point, he concludes: "The highest value of music lies beyond its mere tone colour."[18]

Where might we take this apparent paradox? I would suggest that, rather than undermining the importance of sound and timbre, Pärt's pronouncement serves only to highlight the primacy of the musical content. That in turn rests upon the prior distinction he tacitly makes between pure content (music) and the vehicle of its conveyance (sound). What rescues that distinction from a crude Manichaean spirit-matter dualism is all of the attention that he does in fact devote to sound. In the end, Pärt seems to be saying that music — as notes on a page, which a trained musician can mentally "hear" without the actual movement of air — is at least theoretically distinguishable from its own sound, much like the written word may be distinguished from speech.[19]

To summarize, I would only suggest that, while every composer is concerned with the sonic effect of her or his compositions, Pärt — together with his preferred musicians, ensembles, and producers (notably Manfred Eicher) — has exhibited a particular and meticulous care for sound in his compositions and in the sphere of existential life.

Arvo Pärt

All of this leads inexorably to the last unifying factor of Pärt's music we will identify here, one that might be as self-evident as it is unquantifiable but that cannot be left without mention: the person of Arvo Pärt himself. We have an inkling of that person (or at least of landmark episodes of his life) through the ever-increasing body of accounts of his childhood, personal history, musical training, political pressures, and spiritual journey, before, during, and after the eight-year "gestation period." Much that is not yet written about the events and journeys also conspired to produce this particular person. But in the end, what we have is the human being, whose collective musical, spiritual, and personal characteristics evince a musical language and world. Any artist, of course, is a factor in the creation of her or his art, so it comes down to the improbable task of evaluating the extent of the personal imprint in different cases.

In order to underscore Pärt's identity as a factor in his music, I might at least mention that other composers, not to mention computer algorithms, have produced music that technically adheres to tintinnabuli melody-triad rules, although it would be impossible to mistake these for Pärt. Here too, it may be useful to allow Pärt himself to weigh in, to see how he conceives of the function and identity of the composer in relation to the music. He writes, in his musical diaries:

A composer is a musical instrument and at the same time, a performer on that instrument. The instrument [the composer] has to be in order, to produce sound. One must start with that, not with the music. Through the music, the composer can check whether his instrument is tuned, and to what key it is tuned.[20]

Pärt is drawing attention to the bidirectionality of the relationship between artist and art, composer and music. The composing human subject has to be "in order." The music, in turn, is a measure of that harmonious function. We might suggest expanding that observation beyond a binary inner status of "in tune" and "out of tune," into the music being a conscious self-expression of the composer. We can now see that his pronouncement "Sound is my word" not only draws attention to the sound as a mere or arbitrary vehicle of his inner *logos*. Maybe it means that the sound of his music is true to his *logos*. To shamelessly put words into his mouth, I suggest that he might be saying: "My word — the totality of my self-expression — is latently embodied in the sound of my compositions." This too could be true of many, if not most, great artists (and some mediocre ones as well). But in Pärt's case, it only underscores the fact that — if Neeme Järvi is right, and Pärt's compositions across his entire career are all recognizably his — it is because, in a specific way that we could never fully quantify, they are, in fact, *his*.

That said, I would reiterate the opinion that Pärt's oeuvre is at least as sonically diverse as that of any composer past or present. So much so that anyone identifying a "Pärt sound" with a handful of his early works, as is not uncommon practice among critics, simply hasn't listened to him with sufficient care or comprehensiveness.

Having challenged and investigated the ultimately elusive concept "the sound of Arvo Pärt," we come to our second area of reflection. What is it, within Pärt's compositions, that is making its way to the listener through the vehicle of sound?

On Hearing Pärt

The ideas I have expressed here are based on some relatively objective musicological and biographical details and their relatively subjective effect of potentially constituting a "sound," or at least some kind of central locus of sonic experiences that point back to the composer and his method. The second question — "What are we hearing when we hear Pärt?" — presumes neither complete overlap with nor hermetic isolation from the first. We are here again dealing with a natural admixture of objective and subjective factors. I will focus this section on one root question that I have already shown to play a role in this music: the texts to which it is written. That effectively narrows this constellation of questions to the following root query: What, if anything, of the textual underpinning of Pärt's compositions is "heard" by the listener? As a subset of that question, we must also address why this inquiry is significant.

As for what gets through to the listener, there are as many answers as there are listeners, or perhaps as many answers as there are different instances of aural experience of his music. What are you hearing when you hear Pärt, live (with or without projected supertitles), through headphones (with your eyes closed, or following the text in the CD booklet), in the background as you work on your sculpture (or your physique), on a film soundtrack, or during a television commercial?[21]

Different listeners, at different times, will be more or less attuned to the text of these compositions. People's attention to the text depends partly on the languages they comprehend relative to the text in question. To a native English speaker, *The Deer's Cry* and *The Beatitudes* are likely to carry a greater literary valence than, say, *Kanon pokajanen*, which is in Church Slavonic, a language not fully comprehended even by most Russian speakers. And surely there are many other factors that will dictate, as it were, the effect of the text upon a given listener. Many listeners may be entirely uninterested in the text, either in terms of its content or its role in the music's genesis.

But the significance of the text on the musical product is not limited to the receptivity of the listener. Whether or not the audience cares, the text has made an indelible impress upon the music itself. The music is by and large dictated — in its rhythmic, melodic, and dynamic contours — by its text. Logogenic music is a phenomenon dating back for centuries and across cultures, from the sung recitation of Greek epic poetry, to Gregorian plainchant, to the Jewish practice of scriptural cantillation. Its primary motivation is the comprehensibility of text: If you are singing a text more or less how you say it, with the same kinds of shapes, cadences, and emphases, the public is more likely to understand it. But Pärt's deployment of it is so intentional, and so thorough, that the composer can say, "The words write my music."[22]

The words write his music. That is, they give the music its shape. Several studies have alluded to the syllabic rules that Pärt establishes for a large number of his compositions:[23] long notes for syllables that end a sentence or clause, medium for syllables preceding a comma or in the word preceding a question mark. All other syllables are short. It is sung much like it would be spoken, not in the manner of *recitatif*, but with rhythmic patterns that do not follow fixed meters.

In this way at least, the words are shaping the music and are conveying themselves to the listener. That conveyance is in principle independent of the words' meaning — illustrated still further by those compositions where the music is shaped by the text but the text itself is silent, not even sung. There are at least seven important Pärt compositions where the music is shaped entirely by words — all in Slavonic or Russian, as it happens — that *are not heard*. The motivation of comprehensibility seems to take a gigantic leap into the background because, for all but the most intent and studious listener, the text is not there to be received in any literal sense.

These silent-text compositions lead us deeper into our question of the conveyance and reception of the words. "The words write his music," but that is not only so that they might be received or comprehended in any conventional linguistic sense. There must be something deeper going on for Pärt in this crucially important relationship between text and music.

The next hint of that depth comes with the significance that Pärt accords the scriptural verse so central to the Christian tradition, with implications well beyond it: "In the beginning was the Word" (John 1:1). The first-ever publication of the Arvo Pärt Centre in Estonia is a volume entitled *In Principio: The Word in Arvo Pärt's Music*. At its most pragmatic, musicological level, the book's title could be seen as a restatement of Pärt's *dictum* — "The words write my music"/"In the beginning (of the music) was the word." But in the book's epigraph, Pärt elaborates: "Sound is my word. I am convinced that sound should also speak of what the Word determines. The Word which was in the beginning." In English, where (unlike German) nouns are not capitalized, we notice the significance of his saying "sound is my word" and then following that with saying that sound should be determined by Word — "the Word which was in the beginning" — *in principio*. That Word, in Greek, "Logos," is the "logical" principle according to which the world is made, and in which it has its coherence.[24] The Logos, in Christian understanding, is Jesus Christ.

All of this is to say that the sacred texts to which Pärt sets the vast majority of his compositions are for him more than a set of syllabic shapings and more than a nod to the settings of scriptural verses and masses by the great European classical composers. The texts are at the basis of his faith in God, who creates and saves the world. Pärt's music speaks from the meaning of the words, in sometimes more and sometimes less obvious ways.[25] In *Adam's Lament*, one of Pärt's more programmatic works, music follows meaning quite closely, as one can witness from the opening two and a half minutes. Just after the music/text introduces Adam as the father of humankind, it depicts Paradise with musical lightness and melancholy, and then the estrangement from Paradise, where we hear bitter diminished chords and others that feature alienating intervals — including a minor ninth. Then we hear of the fall, with dissonant chords accompanying a stepwise falling sequence of phrases sung by the male voices. In this case the words are "speaking" not only by conveying their syllabic shape to the melody but through what they are actually telling or narrating to the hearer.

The meaning of the words is obviously important to the composer not only when they tell a story, as they do in the program-music that is *Adam's Lament.* As much as Pärt dodges the image of mystic or monk, and as much as his music is accessible to people of all faiths and none, the unavoidable fact remains: He is a believing, practicing, praying Orthodox Christian. For the purposes of the present reflection, the primary significance of this fact is that the sacred texts at the root of his compositions are of paramount importance and resonance to him personally. In setting classic liturgical prayers and masses he is not merely following in the footsteps of the centuries of Western European composers in their respective settings of scriptural and liturgical texts, though he is certainly doing that. If I may be so bold, he is praying.

Such an assertion is of course impossible to substantiate, yet I would stand by this impression. Through the music, he is directing these words to God, to Christ, to Mary. And he is doing so not abstractly, not as an homage to the liturgical works of Ockeghem, Palestrina, or Bach, but as a reflection of his own intensely held belief. He is doing so in different languages, with their concomitant sonorities. Moreover, he is doing so both in the compositions where the text is sounded and in compositions where it is not.

Moderns and postmoderns have run the gamut of opinions about the meaning (or meaninglessness), the accessibility (or inaccessibility) of authorial intent. With (or without) such theoretical questions in mind, we can ask: Why should we care that he might be praying through this music? Because even as our questions about "hearing Pärt" have to do with the listener, including the question of what (if anything) of the texts finds its way to Pärt's audience, we might profitably turn back to the composer in order to see what he intends to convey and how. Because, effectively, what we are hearing is Pärt's musical translation of sacred text, a translation made more concrete because of the relationship of the composer to the words. Let us stay with this concept of translation a bit longer.

In conversation, Pärt has frequently confirmed his consciousness of this function of his music: "My music is merely a translation of the words." But we must query this use of the word "translation." Because the music is not a translation in the same direct sense that "Je suis un homme" is the translation of "I am a man." Bracketing the semiotic question as to whether and how meaning is reliably conveyed through words at all, when we look at the musical "rendering" of text we are invoking another kind of translation, where words speak through abstraction, as art.

Generally speaking, art is not the uncolored transmission of data; it is the conveyance of meaning by way of intentional creative abstraction. This conveyance is, on the one hand, unreliable in that it may be received so differently by different subjects and in different circumstances. On the other hand, it also has the potential to impress meaning far more powerfully, viscerally, than by the conveyance of cold data.

The discipline of translation studies speaks of "source language" and "target language." And even there we know that there is no direct, unfiltered conveyance of meaning. When we speak of the arts as "translation," our expectations are different. We might do better to speak of "source person" and "target person." The source person is Arvo Pärt. The target person is the listener. And in between the two is the art. The music. The sound.

Back to the vexed question of artistic intent, let us presume that we can have no reliable knowledge of authorial intent and no care as to what it is. But let us be realistic: Are we really that ignorant when it comes to Arvo Pärt? We have established, pretty reliably, that, first, at the source of the music lies the sacred text and, second, that these words carry the deepest possible significance to the artist, such that he is devoted to their meaning, to their divine subject.

The "translation" of these words into music might carry, for some listeners, the sense of the words themselves, addressed as they are to God, Christ, Mary, in a way that would delight Christian believers and estrange nonadherents. But listeners' reactions prove that the musical translation of these heartfelt prayers can carry something bigger than the words' meaning, which is why it speaks so powerfully, and so spiritually, to those with absolutely no buy-in to formal religious belief.

What is happening is this: Pärt renders the specifically Christian sacred text that directly inspires him and gives shape to his music in a form that requires neither piety nor explicit belief. The composer relies on his demonstrated gifts in musically expressing what resides so deeply in his soul. The secular hearer is liable to be moved — less by the Christian specificity of this content than by its appeal to themes shared by religious and nonreligious people alike: the themes of loss and hope, suffering and consolation, death and life, that have widely been observed as at the core of his musical output.[26]

This is why so many people of different walks of life, of all different faiths and of no faith, the spiritual-but-not-religious, find themselves using the same kinds of words to describe Pärt's effect upon them. "Spiritual," "transcendent," "stilling," "bringing me outside myself," "holy," and (a direct quote from a post on the Arvo Pärt Project Facebook page) "I don't believe in God but if I did it would be through this music." Somehow, the intersection of Pärt, his faith, and his relationship to these texts conspires through his compositional skill to produce music evocative of all of these things.

If we were to ask one last time what we are hearing when we hear Arvo Pärt, I would say it is the sound of faith. That sound entails two particularities: the specific faith (Orthodox Christianity) of a specific person (Arvo Pärt). But its translation, into music, into sound, opens out its sphere of access far beyond the particular contours of that confession. What we are hearing, then, is the sound of the composer's faith made accessible through its translation into music.

The music has been received as spiritual, sacred, pure, honest, humble, stilling, sad, and bright by listeners regardless of their faith commitments or lack thereof. It carries these qualities — yes, because of the texts — but more specifically *because of the composer's relationship to them*. He conveys his faith in God, his love of these words and the saints who wrote them, his own experience of living in a broken world, and his experience of sad-brightness. The music's quiet but devastating power derives from his musically praying the texts.

Naturally, his work will register differently with different people, as they "complete" this music in different ways. Each listener completes the act of translation by receiving these sounds into his or her own person and his or her own experience of life's bittersweet vicissitudes. They are receiving a sounding of the sacred — whether "sacred" is taken to mean "divine" or is evocative of a deeply resonant expression of suffering and healing, of broken human life that is in hell but not in despair.

Notes

1. "The hearing of sound is what makes it." Jonathan Sterne, *The Audible Past: Cultural Origins of Sound Reproduction* (Durham, NC: Duke University Press, 2003), 11.

2. "Arvo Pärt" here, and at points elsewhere in this essay, will often function as a synecdoche for "Arvo Pärt's music."

3. Such a claim is impossible to substantiate; it only invites a reconsideration of the common notion of a single "Pärt sound."

4. See Kaire Maimets-Volt, *Mediating the "Idea of One": Arvo Pärt's Pre-Existing Film Music* (Tallinn: Estonian Academy of Music and Theatre, 2009); as well as the distillation/updating of that work in her "Arvo Pärt's Music in Film," *Music and the Moving Image* 6, no. 1 (Spring 2013), 55–71. The listings of films and Pärt compositions there reveal a wider selection that includes several less-well-known works. Still, as for *Spiegel im Spiegel*, the Wikipedia entry for this composition currently lists its use in twenty-five films or trailers between 1996 and 2017, six dance works, and four theater works. https://en.wikipedia.org/wiki/Spiegel_im_Spiegel.

5. As but one example, in the early 2000s the Icelandic band Sigur Rós would frequently play Pärt compositions before they came on stage, in order to put the audi-

ence into the right frame of mind for their own reductive, meditative, quasi-ambient music. For several fascinating examples and stories, see Jeffers Engelhardt, "Perspectives on Arvo Pärt after 1980," in *The Cambridge Companion to Arvo Pärt*, ed. Andrew Shenton (Cambridge: Cambridge University Press, 2012).

6. Cited in Arthur Lubow, "Arvo Pärt: The Sound of Spirit," *New York Times Magazine*, October 15, 2010.

7. "Tintinnabulation is an area I sometimes wander into when I am searching for answers — in my life, my music, my work." Liner notes, *Tabula rasa* (ECM 1275, 1984).

8. See Leopold Brauneiss, "Tintinnabuli: An Introduction," in *Arvo Pärt in Conversation*, ed. Enzo Restagno et al., trans. Robert Crow (Champaign, IL: Dalkey Archive, 2012), 107–62; and Leopold Brauneiss, "Musical Archetypes: The Basic Elements of the Tintinnabuli Style," in *The Cambridge Companion to Arvo Pärt*, ed. Andrew Shenton (Cambridge: Cambridge University Press, 2012), 52–53.

9. See my *Arvo Pärt: Out of Silence* (Yonkers, NY: SVS, 2015), 100–1.

10. Brauneiss, "Tintinnabuli: An Introduction," 109.

11. Perhaps the most celebrated of these transgressions comes within the first-ever tintinnabuli composition, *Für Alina*, with the C# that breaks the triad rule. So striking is that moment that the composer drew a small flower under the note, which is preserved in many reproductions of the score.

12. Epigraph to Hedi Rosma et al., eds., *In Principio: The Word in Arvo Pärt's Music* (Laulasmaa, Estonia: Arvo Pärt Centre, 2014).

13. Opening of Arvo Pärt's short speech at the Vatican in February 2015, as seen in Günter Atteln's film *The Lost Paradise* (Accentus Music, 2015), at the outset of chapter 10 ("Adam's Sin"), minute 39.

14. http://www.arvopart.ee/en/2016/10/one-should-be-careful-about-every-sound-word-act-about-the-programme-of-the-first-tintinnabuli-concert/. Emphasis added.

15. Interview with Björk, 1997 BBC documentary on "Modern Minimalists," available online, on a fluctuating series of YouTube URLs.

16. ECM New Series 1275.

17. As if to emphasize the freedom accorded by "no fixed instrumentation," there is at least one version of *Fratres* on the internet that is performed with, well, *nothing*: It is a "no-input" performance consisting of tones generated by feedback. See Christian Carrière, "Fratres pour console no-input et violon alto," https://vimeo.com/40603278.

18. Cited in the official catalogue of Pärt's work published by Universal Edition. See http://www.universaledition.com/composers-and-works/arvo-part-534.

19. We are approaching the meaning of the memorable quip, once attributed to Mark Twain, "Wagner's music is better than it sounds."

20. See Bouteneff, *Arvo Pärt*, 223.

21. As one remarkable example, the US telecast of the 2018 Super Bowl (viewed by over 100 million people) featured a Jeep commercial set to *Spiegel im Spiegel*.

22. Rosma et al., eds., *In Principio*, 13

23. E.g., Paul Hillier, *Arvo Pärt*, Oxford Studies of Composers (Oxford: Oxford University Press, 1996), 127–28; Bouteneff, *Arvo Pärt*, 73–75.

24. "He is before all things, and in him all things hold together" (Col. 1:17).

25. In invoking "meaning" here, I do not wish to imply that the words of sacred texts have just one fixed meaning. What I hope to convey is that the texts are important to the composer for what they are actually saying (to him) and not only for their syllabic contours.

26. See Bouteneff, *Arvo Pärt*, sec. 3: "Bright Sadness," esp. 139–44.

II

History and Context

3
Sounding Structure, Structured Sound

Toomas Siitan

A close colleague said something to me a few years ago while I was preparing a lengthy lecture series on Arvo Pärt's works for the University of Tartu, and the comment is still fresh in my mind: "It is going to be a real challenge for you to show that pieces Pärt composed during different periods were all done by the same person!" For me, Pärt's works have always formed an integral whole, no matter how dissimilar their soundscapes may be. But my colleague was right: *Am* I capable of putting together a consistent picture for those who hear polar opposites in Pärt's work from different periods? There are contrasts not only of style but also of genre to think about here: Likely there is no other composer of postwar avant-garde music who gained fame for writing serialist pieces who has also written popular children's songs and music for dozens of films and plays (with a preference for animation and puppet theater).

In the canon of scholarly Pärt interpretation (if such a thing already exists), it is common to speak of a watershed moment in the late 1960s and 1970s: On the one side is *Credo* (1968), in which Pärt bids a firm farewell to serialism and collage, and on the other is his invention of the "tintinnabuli" style in 1976. Between these two points were seven or eight years of "silence," or "transition years," as Peter Bouteneff has termed it with much greater accuracy,[1] since it would be entirely incorrect to give the impression that the composer wasn't working intensely during that period. During this time of transition, Pärt composed not only his pivotal Symphony No. 3 (1971) but also the seven-movement oratorio *Laul armastatule* (*Song to the Loved One*, 1971–1973, later withdrawn by Pärt; text by the medieval Georgian poet Shota Rustaveli) and scores for twenty movies. Pärt's emigration to the West in January 1980 — just a handful of years after arriving at his new style of composition — articulates these two contrasting stages of life and composition. This framework is an adequate heuristic, but it can easily become a cliché that prevents one from noticing the clear links connecting these creative periods. For instance, Pärt was indeed especially active in researching early music in the early 1970s, but the traditional titles used for several of his works already composed during his student years and the canon techniques in his Symphony No. 1 ("Polyphonic," in two movements: 1. Canons, 2. Prelude and Fugue, 1963) point not only to the neoclassicism prevalent in Soviet music at the time but also allude to his focus on traditional polyphonic technique, which led him to the use of dodecaphony, or twelve-tone serialism. References to Bach in several of Pärt's works of the 1960s, especially his consistent use of Bach's initials as head-motifs for ten- or twelve-tone rows in a number of his compositions in 1964 (*Diagramme, Missa syllabica, Quintettino, Collage über B-A-C-H*) all seem to lead up to a particular piece: *Wenn Bach Bienen gezüchtet hätte* (1976), which is one of the most complex but least-performed and understudied of Pärt's works from the year he invented the tintinnabuli style.

Pärt's compositional method has been characterized by the search for a solid yet simple structural anchor in his earliest works. At the same time, a central facet of his creative process has been the shaping of a composition's structural concept to complement an expressive, communicative gesture. The significant role of film and theater music in Pärt's compositions from the 1960s is thus in no way coincidental or marginal. What is more, children were often his intended audience: The composer used these genres as a "creative laboratory" for trying out new structural concepts, such as his "serial architecture," which Christopher May has wittily analyzed using the example of Pärt's soundtrack for the 1962 stop-motion film *Väike motoroller* (*The Little Motor Scooter*).[2] Leida Laius (1923–1996) — the first film director with whom Pärt collaborated — was also astounded by the composer's visual, playwright-like instincts when he assisted in the film-editing process.

Allow me to point out a simple, vivid example of the connection between illustrative gesture and organizational structure in one of Pärt's earliest compositions: *Dance of the Ducklings* (*Pardipoegade tants*) from the cycle *Four Easy Dances for Piano: Music for Children's Theater* (*Neli lihtsat tantsu klaverile. Muusika lastenäidenditele*, 1956–1959). In the foundational structure of this children's piano piece, Pärt implemented the very same "musical archetypes" (the diatonic scale and triadic harmony) that Leopold Brauneiss refers to as a starting point for the tintinnabuli style.[3] Here we encounter the mirroring of musical elements that is so common in tintinnabuli, but what is most inventive is how Pärt ties the horizontal to the vertical: Arpeggiated clusters played with the right hand, which resemble a duck waddling over the white keys,

Tanz der Entenküken

Pardipoegade tants * Dance Of The Ducklings

Arvo Pärt
(1956/ 57)

Maßvoll / Parajalt / Measured

Figure 1. *Dance of the Ducklings* (*Pardipoegade tants*) from the cycle *Four Easy Dances for Piano: Music for Children's Theater* (*Neli lihtsat tantsu klaverile. Muusika lastenäidenditele*, Eres 2163). Reproduced with the permission of Eres Edition.

correspond both in pitch content and the ambitus of their movement to a left-hand scale that is twice as slow (Figure 1). Similar proportional canons are frequently found in Pärt's later pieces, where identical material assembles in layers that move in various rhythmic proportions.

Pärt's first twelve-tone scores, composed as a student, did not yet feature a strong, systematic buildup when joining horizontal and vertical dimensions. However, a clear shift occurred in the composer's style immediately after completing his graduate work (Symphony No. 1, titled "Polyphonic"), with the sensationally successful orchestral piece *Perpetuum mobile*, composed in 1963. Here, the more narrative form of his earlier works is replaced by the strict, mathematical, systematic nature of serial technique. Paul Hillier correctly notes that *Perpetuum mobile* "incarnates a compositional archetype,"[4] which becomes characteristic of a great number of Pärt's later works as well — pieces are shaped as a dynamic ebb and flow with a culmination around the golden section, for instance.

After a 1964 performance of *Perpetuum mobile* in Venice, the Italian music critic Giacomo Manzoni wrote that "the composer has not directed useless attention to any kind of popular trend, but has absorbed and generalized the most important experiences of recent years."[5] This opinion is paradoxical: In Soviet Estonia, Pärt could only have had a cursory familiarity with the "important experiences" Manzoni believed he had integrated. Pärt was included in the Warsaw Autumn Festival of Contemporary Music, which was the Eastern bloc's primary (and virtually only) forum for avant-garde music, the same year he composed *Perpetuum mobile* (1963). He relied more on general principles than on "experiences," and the piece met with success as a result of the extreme rationality of the composer's vision as well as its extreme simplicity of structure. The fact that Pärt's manner of compositional thinking was primarily on the structural level in late 1963 is vividly represented by the pruning down of *Perpetuum mobile*'s structure in the choral miniature *Solfeggio*. Instead of a twelve-tone row, the piece features the simplest archetype of what is considered a series: the C-major scale. *Solfeggio* can even be regarded as a "prototintinnabuli" composition: Although triads are not among the base elements, its maximal structural reduction, modes of expression, and harmonic aesthetics, which reconcile the interlaced dissonances into euphony, connect the work to Pärt's later style.

The structural patterns that recur in Pärt's work produce diverse sonic results. Pärt had already experimented in the 1960s with various means of structuring that move from work to work and later shape his compositions in the tintinnabuli style. For example, *Perpetuum mobile*'s great dynamic swell and consequent fragmentation are the result of Pärt's adding and subtracting of structural elements: Layers of progressively shorter durations are added together according to an orderly mathematical pattern to arrive at a culmination of extreme rhythmic complexity, followed by a *diminuendo* that is crafted by a subtraction of layers. The extended swelling to a culmination in the first movement of Pärt's cello concerto *Pro et contra* (1966) has an entirely different musical feel, but the principle behind its structuring is comparable.

Mathematical order helped Pärt distance himself from the subjective and emotional realms of music. By concentrating on basic musical elements, he invented a range of archetypal compositional methods for future use. For instance, Pärt's 1977 double concerto *Tabula rasa*—his first longer tintinnabuli composition—uses mathematical logic in its structuring similar to the kind in *Perpetuum mobile* and *Pro et contra*, although the new musical landscape seems like the polar opposite of his earlier style. Commenting on his 1960s serial compositions in an interview with Enzo Restagno, Pärt states:

> At that time I was convinced that every mathematical formula could be translated into music. I thought that in this way one could create a

more objective and purer kind of music. If I had succeeded by other means in creating a music free of emotions, I would have been able to distance myself from twelve-tone music.[6]

Formulaic writing tends to work the opposite way in tintinnabuli compositions: They no longer seek the musical expression of a formula but instead the formula *for* a musical gesture. In his monograph on Pärt, Hillier quotes from a conversation the composer had with a group of students at the University of Oregon in 1994: "A composition comes as a single gesture which is already, in essence, music. . . . The compositional task is to find the appropriate system for the gesture."[7]

Serialism enabled Pärt to create coherent musical spaces in which melodic and harmonic aspects derive from the same formula. However, the resulting harmonic dimension no longer satisfied him, and during the transition years of the 1970s, Pärt sought to join melody and harmony just as coherently, but in a different way. Prototypes drawn from the history of Western concert music — diatonicism and triadic harmony — are clearly recognizable in Pärt's Symphony No. 3 (1971), where he attempted to roll complexity back into simplicity while adhering to a strictly systematic mode of writing.

Historical Resonances I: Flemish Polyphony

Pärt has discussed his deep interest in fifteenth-century Flemish polyphony during the early 1970s. His first unmistakable reference to music of this era, which he never used in the 1960s collage works, comes in his Symphony No. 3, which features the emblematic use of the so-called Landini cadence characteristic of early fifteenth-century Flemish vocal polyphony. A cornerstone of this style is the fauxbourdon technique, which integrates melodic line and triadic harmony into a systemic whole using a three-part texture. An accompanying line that follows the melody in parallel sixths and octaves is added, while a lower parallel fourth moves along mechanically with the upper line. The result is a chain of 6/3 chords framed by 8/5 chords. The connection to Pärt's tintinnabuli technique is obvious: Both have a rigid bond between melodic line and triadic harmony, and the harmonic plane does not form freely but instead scrupulously follows the melodic movement, adding a vertical dimension.

This technique, which first appeared in European music around 1430 in the compositions of Guillaume Du Fay (1397–1474), is heard explicitly in many of Pärt's pieces: After his Symphony No. 3, it appeared in *Littlemore Tractus* (2000) and *Da pacem Domine* (2004), for instance. As such, this Flemish texture has accompanied Pärt for more than forty years. It is also worth noting that at a concert given by the Estonian early music ensemble Hortus Musicus on October 27, 1976, in Tallinn—the first public performance of the tintinnabuli style—Guillaume Du Fay's *Missa L'homme armé* was performed for the second half of the concert. The choice of style and genre as a companion to Pärt's works cannot be coincidental, especially since Flemish vocal polyphony was rather rare in Hortus Musicus' repertoire. During a public discussion with Pärt held at the Estonian Composers' Festival on June 7, 2013, at St. John's Church, in Tartu, I asked him how conscious he is of this stylistic prototype in his work. The question appeared to confuse him: Apparently, Pärt does not think in the same categories as music scholars and arrived at the connection only through prolonged concentration and reflection.

Word, Text, Music I

Binary opposites exist in music's foundations: sound and silence, movement and stasis, tradition and originality, tonic and dominant, melody and harmony, to name only a few. Pärt endeavors to transcend each of these oppositions in his own way. Schoenberg attempted to transcend the binary of consonance and dissonance as well, but by declaring the equality of all pitches, the outcome was pandissonant music. Already in *Solfeggio* (1963), Pärt sought equality between the seven pitches of the diatonic scale and created a captivating soundscape characterized by dissonant seconds and sevenths continually sounding together, with dissonance now reduced to the qualities in the major scale. Leopold Brauneiss has interpreted the tintinnabuli technique as "the second emancipation of dissonance": it is not avoided or resolved but placed in a balancing triadic context.[8]

Verbal text has a very specific role in Pärt's work. The relationship between text and music can be seen as another binary in the tradition of Western concert music, and Pärt strives to transcend this as well. "The words write my music," he has remarked, which means that Pärt seeks musical structures that are as analogous to texts and speech patterns as possible.[9] German Baroque music displays a similar humility before words, especially in terms of scriptural texts: Heinrich Schütz, for example, called himself "a translator of the story into music," as evidenced in many of his compositions' titles.[10] Baroque composers endeavored to harmonize musical sound and the semantics and rhetorical qualities of a text in order to reveal "the whole meaning" (*die ganze Meinung*) of words, thereby becoming their interpreters. Pärt, on the other hand, proceeds from the formal structure of the word, since he believes in the semantic self-satiety of texts: The full message is perfectly contained within the words, and music merely endeavors to serve and perform, giving the words a sound-based existence.[11]

In *Missa syllabica* (1977), Pärt devised a simple mathematical method for handling text. Initially, the method calculated the number of syllables in a word, but later it accounted for sentence structure (*Cantate Domino canticum novum*, 1977; *De profundis*, 1980), punctuation (*Passio Domini nostri Jesu Christi secundum Joannem*, 1982), and accentuation (*Te Deum*, 1984/85). In *Stabat Mater* (1985), Pärt expanded the textual rules to encompass some of the instrumental *ritornelli*[12] — inversions of the preceding vocal phrases — that subdivide the large-scale work. Thus, the compositional formula for instrumental passages is likewise derived from the text of the corresponding phrase. Pärt repeated this same procedure in several parts of *Berliner Messe* (1990). The possibility to write for instruments on the basis of text was something of a revelation: In September 1985, right around his fiftieth birthday, Pärt drafted *Psalom* (initially without definite instrumentation) on the basis of Psalm 112 (113), and in 1991–92, three pieces comprised a series of instrumental works with "silent texts": *Silouan's Song* (dedicated to one of Pärt's greatest spiritual role models, Archimandrite Sophrony and his brethren), the original version of *Psalom* for string quartet, and *Trisagion*. With *Trisagion*, the text was written into the original score; elsewhere, Pärt merely alluded to the textual associations until he published them in the book *In Principio* (2014).

The instrumental pieces written for silent texts[13] bear particular significance for Pärt: Those expressing personal faith are written in Church Slavonic or Russian (the language of Pärt's religious practice); several pieces bear connections to Archimandrite Sophrony and St. Silouan the Athonite, who are major spiritual figures in Pärt's life; and in most (with the exception of *Lamentate*), a homogenous string sound that mimics the sensitive articulation of liturgical recitation is central to the piece. Initially, the connection to a prayer (the Canon to the Holy Guardian Angel) was the most cryptic element of Pärt's Symphony No. 4 "Los Angeles." The title appeared to be more closely associated with the Los Angeles Philharmonic, which commissioned the piece, and its dedication to the imprisoned Russian oligarch Mikhail Khodorkovsky gave the work a political subtext that was difficult to tie to its musical expression. Pärt, whom the media has long portrayed as a hermetic, monastic figure, had issued strong political statements before, comparing, for example, the October 2006 murder of Anna Politkovskaya to the assassination of Dr. Martin Luther King Jr. and dedicating all performances of his works to the journalist's memory during the 2006–2007 season. Pärt's political statements are not contradictory, because he believes change can only be brought about by prayer that is not perceived externally — just like his "silent text" works. Pärt's politically charged protest in *Credo* was similar: He sought egress not through opposition to violence but through a change within oneself. The lesson from Jesus's Sermon on the Mount, which Pärt chose for the culmination of *Credo*—"But I say unto you, That ye resist not evil"—remained unintelligible to its Estonian audience at the time, being in Latin, and was of course not translated in the program. This text—the very first that Pärt used in a piece after his student years—also remained "silent," in a way.

Historical Resonances II: Conrad Beissel

It is difficult to find authored examples of music structured around the accents and number of syllables in words, such as is common in Pärt's text-based compositions; however, the connection is prevalent in liturgical chanting. One surprising parallel comes from North America—from the German religious émigrés in Lancaster County, Pennsylvania. In 1732, German-born Conrad Beissel (1691–1768) founded the Ephrata Community of pietistic religious refugees—der Lager der Einsamen (the Camp of the Solitaries, or the Ephrata Cloister of the Seventh-Day Baptists), a Protestant community of mystical Christian devotion and practice.[14] Beissel, who remained the community's spiritual leader until his death, was an outstanding amateur musician, composing over one thousand hymns and appointing himself Kapellmeister of the Ephrata Cloister.[15] Entirely self-taught, Beissel created a singular system of harmony. The historical accounts of visitors to the Ephrata community tell of the peculiar sweetness, strange beauty, impressive cadence, and even angelic or celestial quality of its hymns and choral pieces.[16]

Thomas Mann describes Beissel's activities with remarkable historical precision in chapter 8 of his novel *Doctor Faustus*. He details Beissel's harmonic system, which was based on rational chord tables, and suggests that Beissel can be viewed as a forerunner of twentieth-century serialism. His simple, purely text-based rhythmic system has also attracted attention. Accented syllables were marked with longer notes and unstressed syllables with shorter notes, without there being a definite relation between the durations. Even so, the basis for Beissel's compositional method was a simple melodic teaching that divided notes into "masters" and "servants": "Having decided to regard the common chord as the melodic center of any given key, he called the 'masters' the notes belonging to this chord, and the rest of the scale 'servants.' And those syllables of a text upon which the accent lay had always to be presented by a 'master,' the unaccented by a 'servant.'"[17] This entire description resembles to a remarkable degree the principles of musical composition, and especially the approach to text, that have characterized Pärt's works for forty years.

Word, Text, Music II

The Word is sacred: This conviction has shaped the relationship between text and music for both Beissel and Pärt. Pärt included the opening verses of the Gospel of John (1:1–14) in *In principio*, his masterpiece from 2003, and one of his most oft-quoted statements puts this in context: "Sound is my word. I am convinced that sound should also speak of what the Word determines. The Word, which was in the beginning."[18] Perhaps it would not be misconstrued simply to say: "Sound *is* the Word." And as in the Christian tradition, if Jesus is understood as God's Word, God's perfect self-expression,[19] then we may perhaps even paraphrase in this way: "Music is the self-expression of the Word." In this sense, Pärt seeks a hypostatic union between the Word/*logos* and music as it is understood in Christian theological tradition, as well as in the way Nicholas of Cusa understands the union between Man and universe in his renowned *De docta ignorantia* (*On Learned Ignorance*) (1440).

Postwar avant-garde music has often been accused of intellectualism, set against a "mourning for the loss of naïveté adapted and exploited by mass culture," as Theodor Adorno puts it.[20] However, a similar sense of mourning arises from the sharpened perception of a lost natural feeling of unity and from the breaking of tradition. Just as how "coinciding opposites" have shaped the aesthetics of Pärt's compositions on various levels,[21] so has the composer consistently sought a connection between the intellectual and the naïve — and here, naïveté should be understood as "pure," "natural," "complete," and "not self-reflective," in Friedrich Schiller's sense in *On Naïve and Sentimental Poetry*,[22] thereby contradicting judgments that derive from reason and understanding. These categories appear as opposites in Pärt's early work, where extremely rational forms of musical expression stand side by side with music written for children, or in his Symphony No. 2 (1966), where the anguish of dodecaphony contrasts with a starkly emotional passage from Tchaikovsky's *Sweet Daydream* — a piece published in his classic 1878 album of children's piano pieces. Opposites were united in the "complex simplicity" of Pärt's tintinnabuli technique.

Throughout his compositions, Pärt seeks to create music congruent with the laws of harmony and a Pythagorean notion of the cosmos's numerical structure. Since many of his own aphorisms are quoted with excessive frequency but still more of his thoughts are preserved only in private notebooks, allow me to conclude with a quotation from Sofia Gubaidulina, Pärt's peer in terms of generation and artistic ideas. Their respective modes of compositional expression, both of which matured under the conditions of Soviet oppression, are dissimilar in nature but nevertheless linked by a similar understanding of music's role and deeper meaning:

Many have said that music is the most spiritual form of art there is. I've wondered: why is this, exactly? The thing is that we musicians deal with a sound, with a single tone. And that in and of itself contains the pattern of the universe. The fundamental laws of everything that exists are in a constant state of pushing and pulling. And that exists in our material, in sound, which is vibration. It isn't a metaphor or a symbol — it's a fact. It contains the divine pattern of all existence. No other form of art has this kind of material.[23]

When a composer like Arvo Pärt seeks the most elementary and coherent sound structures, he is not striving for simplicity or widespread communication in his music. Rather, what he is pursuing is *unio mystica* — unity with the self-expression of the universe.

Notes

1. Peter C. Bouteneff, *Arvo Pärt: Out of Silence* (Yonkers, NY: SVS, 2015), 86.

2. Christopher J. May, "System, Gesture, Rhetoric: Contexts for Rethinking Tintinnabuli in the Music of Arvo Pärt," D.Phil. thesis, University of Oxford, 2016, 148–50.

3. Leopold Brauneiss, "Musical Archetypes: The Basic Elements of the Tintinnabuli Style," in *The Cambridge Companion to Arvo Pärt*, ed. Andrew Shenton (New York: Cambridge University Press, 2012), 49–75.

4. Paul Hillier, *Arvo Pärt* (Oxford: Oxford University Press, 1997), 46.

5. Leo Normet, "The Beginning Is Silence," *Teater. Muusika. Kino.* 6, no. 7 (1988): 19–31.

6. Enzo Restagno et al., *Arvo Pärt in Conversation*, trans. Robert Crow (Champaign, IL: Dalkey Archive, 2012), 15.

7. Hillier, *Arvo Pärt*, 201.

8. Leopold Brauneiss, "Vähem on rohkem: vabatahtlik enesepiirang tintinnabulistiilis, selle olemus ja tähendus esteetilises, kultuurilises ja vaimses kontekstis," in *Arvo Pärdi tintinnabuli-stiil: arhetüübid ja geomeetria*, ed. Saale Kareda (Laulasmaa: Arvo Pärt Centre, 2017), 39.

9. Toomas Siitan, "Introduction," in *In Principio: The Word in Arvo Pärt's Music*, ed. Hedi Rosma et al. (Laulasmaa: Arvo Pärt Centre, 2014), 13.

10. E.g., Heinrich Schütz, *Historia der frölichen und Siegreichen Aufferstehung unsers einigen Erlösers und Seligmachers Jesu Christi . . . in die Music übersetzet durch Henrich Schützen* (Dresden: Gimel Bergen, 1623).

11. Siitan, "Introduction," 11.

12. E.g., in bars 128–132, 228–232, 245–248, etc.

13. *Psalom* (1985/1991), *Silouan's Song* (1991), *Trisagion* (1992), *Orient & Occident*

(2000), *Lamentate* (2002), *Für Lennart in memoriam* (2006), Symphony No. 4 "Los Angeles" (2008).

14. Jeff Bach, *Voices of the Turtledoves: The Sacred World of Ephrata* (Göttingen: Vandenhoeck & Ruprecht, 2003), 3–4.

15. Lloyd G. Blakely, "Johann Conrad Beissel and Music of the Ephrata Cloister," *Journal of Research in Music Education* 15, no. 2 (Summer 1967): 120–38.

16. Julius F. Sachse, *The Music of the Ephrata Cloister; also Conrad Beissel's Treatise on Music as Set Forth in a Preface to the "Turtel taube" of 1747, Amplified with Fac-simile Reproductions of Parts of the Text and Some Original Ephrata Music of the Weyrauchs hügel, 1739; Rosen und lilien, 1745; Turtel taube, 1747; Choral buch, 1754, etc.,* (Lancaster, PA: Printed for the author, 1903), 11.

17. Thomas Mann, *Doctor Faustus: The Life of the German Composer Adrian Leverkühn as Told by a Friend*, trans. H. T. Lowe-Porter (London: David Campbell, 1992), 64.

18. Rosma et al., eds., *In Principio*, 5.

19. Bouteneff, *Arvo Pärt*, 116.

20. Theodor W. Adorno, *Philosophy of New Music*, trans. Robert Hullot-Kentor (Minneapolis: University of Minnesota Press, 2006), 15.

21. Leopold Brauneiss, "The Unification of Opposites: The Tintinnabuli Style in the Light of the Philosophy of Nicolaus Cusanus," *Music & Literature* 1 (2012): 53–60.

22. Friedrich Schiller, "Über naive und sentimentalische Dichtung," *Die Horen* (1795/96).

23. Sofia Gubaidulina, interview with Timo Steiner (18:40–20:30), *MI*, Estonian Public Broadcasting (ETV), October 20, 2016, http://arhiiv.err.ee/guid/201610201153 00201000300112290E2BA238B440000000438oBooooooDoF028004.

4

Colorful Dreams

Exploring Pärt's Soviet Film Music

Christopher J. May

By now, the "life and works" narrative of Arvo Pärt's Soviet years might seem well settled. It was consolidated by Paul Hillier in 1997,[1] and the critical literature has rehearsed it many times since: the twelve-tone controversy around *Nekrolog*, the move to pastiche, the *Credo* crisis, the "silent period," the tintinnabuli revelation, the eventual emigration in 1980. It is a comfortable procession of key pieces and events, and it closely adheres to Pärt's official list of concert compositions. This chapter, however, claims that that narrative is incomplete. While in the Soviet Union, Pärt produced a great deal of music that was, until 2019, withheld from his authorized works list — notably, a corpus of original film scores.[2] Pärt received his first film commission while still a student at the Tallinn Conservatory, and over the next seventeen years he went on to fulfill dozens more. At times this work was his main source of income, especially in the early 1970s (when his offerings for the concert hall slowed to a trickle). Film scores, in fact, were a prominent and consistent presence in Pärt's Soviet-era output, including during the formative years of tintinnabuli.

One might expect that Pärt scholars would have something to say about this substantial, diverse body of work. For the most part, however, one would search the existing literature in vain. Indeed, it has become uncontroversial to dismiss Pärt's film scores as, collectively, either insignificant or irrelevant. Here too Hillier set the precedent, declaring the film scores to be "wage music" that had "practically nothing to do with [Pärt's] real work as a composer."[3] More recent writings confirm a broad consensus that this output stands wholly apart from the prevailing life-works narrative associated with Pärt's Soviet years.[4] Certainly, no sustained attempt has yet been made to link Pärt's film work with the emergence of tintinnabuli. On most critical accounts, accordingly, these two spheres of Pärt's compositional activity took place in totally unrelated environments and existed in mutually exclusive sonic worlds. It is as though the film scores had been written by somebody else.

I

Recounting Pärt's compositional history without his film oeuvre involves, at least implicitly, the assumption that these scores do not meaningfully add to an overall understanding of his music. This chapter challenges that position. Here I confront the habit of disengaging from Pärt's film music, and I examine what might underlie decisions to include and exclude specific pieces and facts when building critical narratives around him. Via case studies, I also seek to establish the study of Pärt's film scores as a highly productive exercise, pointing out in particular their capacity to enrich our knowledge of the origins and development of tintinnabuli. In that sense, this chapter stands alongside Kevin Karnes's contribution, offering a reminder that among the "forgotten sounds" of Pärt's world must be counted much music by Pärt himself. However, since this topic is so minimally documented in Anglophone scholarship, it is appropriate to begin with a short general overview of Pärt's film output.

When researching this repertory in 2015, I at first had difficulty fixing even the most basic information about it. Fortunately, Kaire Maimets had already assembled and published a proposed filmography for Pärt's Soviet years as part of her 2004 article on his score for Leida Laius's film *Ukuaru* (the source of the *Ukuaru valss*).[5] This list proved an invaluable starting point, and I crosschecked it with many other sources, including catalogs of the Estonian Film Institute and Estonian Public Broadcasting, the Estonian Composers' Union reports of its members' activities, and the credits of the films themselves (wherever I could obtain them; some are more easily found than others). I also used secondary databases to corroborate attribution details. Perhaps inevitably, this process threw up a host of small difficulties to do with consistency, reliability, and completeness among the various sources. I resolved these as best I could, and the results appear in Table 1, a catalog that both expands upon Maimets's original and revises some of its particulars. This listing is, I believe, the first of its kind to appear in English.

Some thirty-four original film scores may be safely attributed to Pärt between 1962 and 1979 — about one every six months. A further seven ascriptions are more dubious. Two production companies were active in Estonia in this period — Tallinnfilm and Eesti Telefilm — and Pärt worked with both, often in repeat collaborations with directors. Occasionally, his films also received a wider distribution within the Soviet Union. He scored across a variety of

Table 1: Proposed tabulation of Pärt's original Soviet film scores

No.	Year	Title	Director	Studio	Genre	Length (mins)	Language	Estonian Film Archive catalog number
1	1962	*Õhtust hommikuni* (From Evening to Morning)	Leida Laius	Tallinnfilm	feature (*mängufilm*)	37	Estonian, Russian, German	1826
2	1962	*Väike motoroller* (The Little Motor-Scooter)	Heino Pars	Tallinnfilm	puppet-animation (*nukufilm*)	10	Estonian	4774
3	1963	*Just nii!* (Just So!)	Elbert Tuganov	Tallinnfilm	puppet-animation	10	none	4929
4	1964	*Operaator Kõps seeneriigis* (Cameraman Kõps in the Land of Mushrooms)	Heino Pars	Tallinnfilm	puppet-animation	21	Estonian	4919
5	1964	*Viimne korstnapühkija* (The Last Chimney-Sweep)	Elbert Tuganov	Tallinnfilm	puppet-animation	11	Estonian	4775
6	1964	*Evald Okas* (Evald Okas)	Virve Aruoja	Eesti Telefilm	documentary (*dokumentaalfilm*)	19	Russian, Estonian	1797
7	1965	*Mäeküla piimamees* (The Milkman of the Manor)	Leida Laius	Tallinnfilm	feature	88	Estonian	1833
8	1965	*Hiirejaht* (Mouse-Hunt)	Elbert Tuganov	Tallinnfilm	puppet-animation	10	unknown	4938
9	1965	*Operaator Kõps marjametsas* (Cameraman Kõps in the Berry Forest)	Heino Pars	Tallinnfilm	puppet-animation	20	Estonian	4920
10	1966	*Operaator Kõps üksikul saarel* (Cameraman Kõps on a Desolate Island)	Heino Pars	Tallinnfilm	puppet-animation	19	Estonian	4918

11	1967	*Kurepoeg* (The Young Stork)	Elbert Tuganov	Tallinnfilm	puppet-animation	19	none	5423
12	1968	*Operaator Kõps kiviriigis* (Cameraman Kõps in the Land of Stones)	Heino Pars	Tallinnfilm	puppet-animation	19	Estonian	4917
13	1969	*Enderby valge maa* (Enderby — White Land)	Andres Sööt, Mati Kask	Tallinnfilm	documentary	37	Estonian	2061
14	1970	*Aatomik* (Atom-Boy)	Elbert Tuganov	Tallinnfilm	puppet-animation	9	Estonian	5234
15	1970	*Aatomik ja jõmmid* (Atom-Boy and the Thugs)	Elbert Tuganov	Tallinnfilm	puppet-animation	10	none	5231
16	1970	*Mis? Kes? Kus?* (What? Who? Where?)	Heinz Valk	Tallinnfilm	puppet-animation	8	Estonian	5419
17	1970	*Jäärik* (The Ice Realm)	Andres Sööt, Mati Kask	Tallinnfilm	popular science (*populaarteaduslik film*)	14	Estonian	4882
18	1971	*Putukate suvemängud* (The Insects' Summer Games)	Heino Pars	Tallinnfilm	puppet-animation	19	Estonian	5260
19	1972	*Helin* (Ringing)	Olav Neuland	Eesti Telefilm	documentary	39	Estonian	
20	1972	*Inimeselt inimesele* (From Person to Person)	Rein Maran	Eesti Telefilm	documentary	30	Estonian	
21	1972	*Otsin luiteid* (Searching for Dunes)	Hans Roosipuu	Tallinnfilm	documentary	10	Estonian	2057
22	1973	*Ukuaru* (Ukuaru)	Leida Laius	Tallinnfilm	feature	85	Estonian	4093
23	1973	*Pallid* (Balls)	Heino Pars	Tallinnfilm	puppet-animation	8	Estonian	5391

(continued)

Table 1: (continued)

No.	Year	Title	Director	Studio	Genre	Length (mins)	Language	Estonian Film Archive catalog number
24	1973	*Veealused* (Sea-creatures)	Heino Pars	Tallinnfilm	puppet-animation	16	Estonian	5264
25	1973	*Sellised lood* (Such Stories)	Ants Kivirähk	Eesti Telefilm	cartoon (*joonisfilm*)	14	none	
26	1973	*Värvipliiatsid* (Colored Pencils)	Avo Paistik	Tallinnfilm	cartoon	10	none	5265
27	1973	*Kaugsõit* (The Long Journey)	Andres Sööt, Mati Kask	Eesti Telefilm	documentary	30	Russian	1860
28	1974	*Värvilised unenäod* (Colorful Dreams)	Virve Aruoja, Jaan Tooming	Tallinnfilm	feature	60	Estonian	5173
29	1974	*Õed* (Sisters)	Elbert Tuganov	Tallinnfilm	puppet-animation	8	none	5448
30	1974	*Täheke* (Starlet)	Avo Paistik	Tallinnfilm	cartoon	8	none	5442
31	1975	*Briljandid proletariaadi diktatuurile* (Diamonds for the Dictatorship of the Proletariat)	Grigori Kromanov	Tallinnfilm	feature	151	Russian	5927
32	1975	*Looduse hääled* (Voices of Nature)	Rein Maran	Tallinnfilm	popular science	unknown	Estonian	4733
33	1978	*Jäljed lumel* (Footprints in the Snow)	Leida Laius	Tallinnfilm	documentary	26	Estonian	2980
34	1979	*Navigator Pirx* (Pilot Pirx's Test)	Marek Piestrak	Tallinnfilm and Zespoly Filmowe	feature	94	Polish, Russian	6021

Doubtful attributions and misattributions					
No.	Year	Title	Director	Studio	Genre	Reason for doubt
35	1964	*Inimesed arsenalis* (People in the Arsenal)	unknown	unknown	unknown	Only source is Estonian Composers' Union reports of members' activities. Listing appears together with *Evald Okas* but apart from other film scores.
36	1965	*Reekviem* (Requiem)	Jaak Mamers	Eesti Telefilm	documentary	Only source is Estonian Film Institute website. No corroborating attribution yet found.
37	1966	*Reportaaž telefoniraamatu järgi* (Journalism by the Telephone Book)	Virve Koppel, Mati Põldre	Eesti Telefilm	documentary	Not original: reuse of Pärt's early piano works, other music by Kuldar Sink.
38	1971	*Inspiratsioon* (Inspiration)	Valdur Himbek	Tallinnfilm	documentary	Not original: reuse of *Collage* and *Musica Sillabica*.
39	1972	*Kolm portreed* (Three Portraits)	Olav Neuland	Eesti Telefilm	documentary	Almost certainly an alternative title for *Helin*.
40	1973	*Kapten* (Captain)	Olav Neuland	Eesti Telefilm	documentary	Almost certainly a reference to part 2 of *Helin*.
41	1976	Peace	Boshra Abo-Saif	unknown	unknown	Appears to be an American student film, hence not an original score.

Shading indicates that manuscript material for the film is held at the Estonian Theater and Music Museum (ETMM).

Work at the Arvo Pärt Centre was not possible and work at the archives of the production companies was not practicable at the time this research was carried out.

genres, from full-length features and documentaries to shorter children's puppet animations, and some of these films are well known in Estonia to this day.
There is a daunting amount of music to be examined here, and the variety
of musical styles employed by Pärt across the full corpus is most striking and
intriguing (why, for example, does Mozartian pastiche feature in so many of
his film scores from the early 1970s?). Happily, prospective researchers are not
left to rely solely on their aural impressions and transcription skills, since the
Estonian Theatre and Music Museum in Tallinn (hereafter ETMM) holds
original manuscript material for fifteen of the thirty-four scores. While these
documents can be patchy and are sometimes difficult to read and interpret,
they are nonetheless a crucial source of information for some of the arguments
put forward here to link the cinematic Pärt with the Pärt of the concert hall.

To engage fully with this body of music would require a dedicated project.
We know almost nothing about matters of potentially great significance: the
reception of these films and their scores, the expectations of the commissions,
and Pärt's own designs for the scores (as opposed to what, following directors'
decisions and censors' scrutiny, made the final cut). Comprehensive further
research would involve sources such as documents kept by the creative unions
in Soviet Estonia, contemporary cultural newspapers, and the soundtrack recording sessions (a number of which survive),[6] as well as some oral history
and transcription. All experiences of dealing with the historical records of
Soviet-era art suggest that such materials might well be riddled with ambiguity. Ambiguity can, of course, be critically fertile. It is nonetheless apparent
that producing a nuanced account of Pärt's work as a film composer would be
a difficult undertaking.

Accordingly, this chapter's goals are more limited. I want to show that Pärt's
film and concert music are closely interrelated, and I do this primarily through
examples chosen to illustrate the extent to which film work was a platform for
Pärt's technical development throughout his Soviet years. I also consider how
the inclusion of the film scores within Pärt discourse might affect how we understand the origins, and origin myths, of tintinnabuli. Perhaps most importantly, I am concerned with establishing an appropriate critical framework —
since none currently exists — within which to situate all these inquiries. If we
elect to stop simply ignoring this music, then how are we to approach it in an
aesthetic, epistemological, or analytical sense? It is this question, by no means
an easily answered one, to which I turn first.

II

Putting aside the readily apparent — and clearly significant — practical issues
of accessibility and language, I can suggest three major reasons why Anglo-

American scholars have consistently declined to engage with Pärt's film scores. The first is the composer's own attitude toward this music. For many years, Pärt did not officially recognize his film output. He dislikes talking about it and has actively discouraged research on it.[7] More than this, Pärt has described his Soviet film music in ways that have seemingly licensed critics to reject, in advance, any possibility of its artistic validity, technical significance, or intertextual relevance. For instance, he has associated it with a culture of unsympathetic pragmatism, mentioning the tight deadlines and budgets of the Soviet planned economy, which often led to hastily written scores and rushed recordings. Pärt has also stated that the censoring of finished films would often totally disregard any effects on the music, which was "cut like some sausage."[8] These unpromising images of creative disenfranchisement tally neatly, of course, with Hillier's "wage music" remark.

Certainly it is a composer's right to withdraw music. But does that step oblige critics to ignore the music's potential discursive relevance? Is it scholars or composers who have ultimate rights over critical knowledge and over the processes by which historical facts are made into interpretive narratives? And how far can one realistically press, if one's project depends on the ongoing goodwill of a composer's circle? What constitutes responsible and ethical scholarship here is clearly a subjective matter. The film music is by no means Pärt's only unofficial critical taboo, and difficult issues of scholarly distance and objectivity have colored the writing of most major Pärt researchers.[9] While my own work has certainly involved compromise, I incline to the view that the worse option is to eliminate interpretive possibilities before exploring them. To accept Pärt's disavowal of his film scores unquestioningly is still to make a critical choice: the choice to defer to Pärt as a final interpreter of his own music. Yet he is not typically regarded in that way — nor, as Andrew Shenton rightly remarks, does he even see himself as that kind of authority.[10] Hence the accession to Pärt's preferences regarding his penumbral works is always a selective one. It embraces, I suggest, a ready-made excuse to suspend inquiries threatening to come into tension with a sensitive and perhaps personally uncomfortable chapter of Pärt's history. This is all understandable. But for a scholar determined to remain independent, it ought not necessarily be decisive.

In any event, Pärt has also said plenty to suggest that scoring films did play a real role in his compositional development. A key factor here was an institutional loophole: Film music fell outside the Composers' Union remit and thus avoided scrutiny by professional colleagues. In effect, this meant that there was no music-specific censor and that music unthinkable for the concert hall could nonetheless be written for films. Pärt has described this more than once as an opportunity for freedom of personal expression, and Schmelz has indicated how the most controversial members of Pärt's generation — Schnittke, Denisov, Gubaidulina, Volkonsky — all at some point turned to the relatively permissive conditions of film composition. It would hence seem natural to approach Pärt's film work as a window into his serious compositional interests. This also makes sense, I suggest, within a Soviet context where, as Pärt noted to Enzo Restagno, film scoring was a standard branch of compositional activity, not a separate area of specialization.[11] Tatiana Egorova's study of Soviet film music confirms the number of "outstanding composers" — Shostakovich, Prokofiev, Sviridov — who were involved in the genre.[12] One might expect such a culture to promote the transference of skills and ideas between film and art music, and Pärt has in fact accepted the likelihood of such connections: "This should leave a mark after all. It is somewhere there but it needs to be found."[13] All of this means that any critical narrative intent on discounting or quarantining Pärt's film music should, if it is to be credible, first contend with a musical culture in which film scoring was both a highly respected compositional pursuit and a logical vehicle for experimentalist composers to hone their craft and technique.

Pärt's personal discomfort may be related to another likely reason for his film scores' neglect: the critical phenomenon labeled "binary socialism" by the anthropologist Alexei Yurchak. This term refers to a habitual use of binary oppositions to describe Soviet life, with the most important including oppression versus resistance, state versus people, official culture versus counterculture, public versus private self, and corruption versus morality. Yurchak associates binary socialism with many accounts of Soviet culture produced both in the West and (since 1991) in the former Soviet Union, and he attributes the flourishing of these reductive labels to the "antisocialist, nonsocialist and postsocialist political, moral, and cultural agendas and truths" that dominate the contexts within which most critical knowledge about socialism is produced. Yurchak's own contention, by contrast, is that most day-to-day experiences of Soviet life did not fit within binary socialist categories. Rather, and without denying the presence of repression and corruption within the Soviet state, he draws attention to social identities and practices that reveal the normality of fluid and multiply signifying relationships among individuals, groups, and an ideologically structured cultural discourse.[14] Authors such as Levon Hakobian have made similar arguments in the context of music criticism specifically. Hakobian's book on Soviet music opens with a withering rejection of stereotypical images invoked in the West to depict Soviet life, and he goes on — targeting Shostakovich studies in particular — to give short shrift to the notion of a fixed artistic typology of "corrupt elites, courageous dissidents, and . . . silent sterile conformists."[15] While binary models of socialism tend to embed a presumption that political position taking was a primary plane of meaning for Soviet citizens engaged in creative practices, both Yurchak and Hakobian are explicit that this need not necessarily be so at all.

Despite such forthright literature, Pärt's Soviet-era scores are not infrequently approached in a manner that would fall within the scope of Yurchak's criticism. The compositions that attracted official opprobrium — *Nekrolog*, *Credo* — typically receive plenty of interest: Indeed, sources going back to the 1980s have tried hard to maximize Pärt's anti-Soviet credentials by dwelling on his run-ins with the cultural authorities.[16] Meanwhile, any music that seems ideologically correct is overlooked, a good example being Hillier's treatment of *Maailma samm* (*Stride of the World*), Pärt's socialist realist oratorio from 1961. Hillier is at pains to dismiss this work as artistically inauthentic: He finds it "heavy with forced optimism and artificially inseminated international benevolence" and calls it an instance of the "tuneful optimism that earned Soviet prizes." He proceeds to place *Maailma samm* outside Pärt's "true path" — that path, evidently, being the path leading not to prizes but to scandal and rebuke.[17] In fact, however, *Maailma samm* contains exactly the same "bourgeois" technique, namely, twelve-tone music, that caused all the trouble with *Nekrolog*. Accordingly, this passage reveals considerably more about Hillier's critical perspective than about Pärt's music. The idea that Pärt must have worn a mask when writing *Maailma samm* and that he must hence be distanced from it in order to preserve his artistic integrity reflects the basic moral logic of "binary socialism."[18]

Challenges to such bowdlerizing criticism have been rather rare. It seems probable, therefore, that anxieties about the "true Pärt" and the like have motivated selective dealing not only with *Maailma samm* but also with other Pärt scores on the periphery of critical awareness. These anxieties, though reinforced by the composer himself, are in my view somewhat misdirected. By stigmatizing certain works, they impede interrogation of the full range of meanings — less politicized ones, perhaps — that might be caught up in the compositional choices and stylistic variety on display throughout Pärt's Soviet music. As Peter Schmelz rightly argues in his study of "unofficial" music from the Thaw, creative activity *in relation to* prevailing cultural ideologies need not necessarily signify either an opposition to or an affirmation of those ideologies.[19] This echoes the language of Yurchak and Hakobian, as well as writing by scholars such as Katherine Verdery, who has described the phenomenon of "public" and "private" selves under socialism in terms that signal interdependence more strongly than conflict.[20] I suggest, accordingly, that critical encounters with Pärt's film music are most likely to be productive if they acknowledge, as "binary socialism" does not, the polyvalent and often ambiguous conditions within which professional Soviet composers worked. Here I will argue that two of his earliest scores for the cinema may be plausibly interpreted to signify both "within" and "beyond" the prescriptions of Soviet cultural ideology.

Lastly, decisions to minimize Pärt's film music may stem from an entrenched habit of conceiving his music in discrete *periods*. Critical writing often articulates an archetypal narrative under which Pärt, the protagonist, first experiences an artistic and personal crisis in the 1960s, involving turbulent encounters with modernist styles and Soviet ideology. He then undergoes a period of self-discovery during his so-called silent period of the early 1970s, and finally achieves a resolution with the "birth of tintinnabuli" in 1976. This basic outline underpinned much of the mythmaking that took place in Pärt reception during the 1980s and 1990s (for example, it is glimpsed in the prominent writings of Wolfgang Sandner). In a slightly modified version, it organizes the composer's biography on the website of the Arvo Pärt Centre (hereafter APC). It was also adopted in Hillier's highly influential monograph, in which *Credo* is explicitly made emblematic of culminating crisis and *Für Alina* of creative rebirth.[21] A more recent illustration of its discursive normalization came in 2015, when Donald Macleod featured Pärt on BBC Radio 3's *Composer of the Week*. Macleod spoke of "two distinct periods" in Pärt's music, divided by extended contemplation; he described tintinnabuli, meanwhile, as emerging from nothing after a prolonged search for a mysterious inner source, an act of "coming through on the other side."[22]

This arc of crisis, self-discovery, and resolution explains much about the composer and his music with validity and insight. However, it is certainly not neutral, and it freights the tintinnabuli concept in particular in two important ways. First, it is a narrative of rupture. By casting tintinnabuli as a creative rebirth from nothing, Hillier, Macleod, and the others sustain the idea of its total separability from Pärt's earlier music. Tintinnabuli is imagined in opposition to various "others" (dodecaphony, collage, years of silence), and segregation of Pärt's repertory swiftly ensues. The corollary — on display in writings by Leopold Brauneiss, among others — is that any evidence of stylistic or aesthetic continuity between the periods is likely to be devalued or overlooked.[23] Second, the archetypal narrative is highly teleological, promoting an evaluative discourse that understands tintinnabuli as Pärt's "true" style: his best, most authentic, and most worthwhile music. The 2012 *Cambridge Companion to Arvo Pärt* illustrates this phenomenon well: Taking tintinnabuli as virtually its sole field of inquiry, its implicit message is that Pärt studies might readily dispense with all other music by the composer.[24] Such entrenchments of tintinnabuli's discursive privilege go hand in hand with the canon-forming influence of Pärt's preferred record label, ECM, which has historically shown little interest in releasing his pre-tintinnabuli works.

It is all too clear how easily Pärt's film output fades from view within such a framework. His cinematic scores simply do not fit, continuing as they do through all three "periods" and harnessing diverse and unpredictable styles along the way. Ultimately, therefore, it is simpler just to regard them as one more of tintinnabuli's relatively unimportant "others." Yet any presumption of rupture between these two Pärtian oeuvres — tintinnabuli and film — is readily falsified. In fact, clear intertextual connections appear in the very first tintinnabuli work to be completed in 1976, *Sarah Was Ninety Years Old.*[25] As Karnes has noted, important pitch material from *Sarah* had already appeared two years previously in Pärt's score to an Estonian feature film called *Värvilised unenäod (Colorful Dreams).*[26] Having debuted it in the film score and refined it for *Sarah*, Pärt then reused this same material in at least two subsequent tintinnabuli works: his so-called *Italian Concerto* (now withdrawn) and his choral motet *The Woman with the Alabaster Box.* It would be misleading to suggest that Pärt's film scores are riddled with such obvious overlaps. Rather, I cite *Värvilised unenäod* here simply to show that there do exist undeniable links between the tintinnabuli and film outputs — and consequently, that there is critical risk in narrating one in a way that neglects the other.

Once examined, then, the major reasons for justifying the critical exclusion of Pärt's film music — the composer's discouragement, binary socialism, the archetypal narrative — do not seem especially robust. To the contrary, there is good reason to suspect that we have missed much by not yet taking this output seriously. Ultimately, and somewhat ironically, it was Pärt himself who helped establish the critical framework for my own inquiries by affirming to me (through the APC) the status of his film music as a "creative laboratory."[27] This metaphor usefully leaves room for a feature of film music on which Pärt has always insisted: its ontological distinctiveness (film scores are not "works" and require a distinct and particular set of compositional skills).[28] Accordingly, Pärt's phrase guides a critical approach that is unafraid to acknowledge permeability between his film corpus and wider oeuvre but also recognizes that specifically filmic modes of musical creativity and signification do compromise the validity of direct, like-for-like comparisons with his authorized list of works. It is on this basis that I apply the "creative laboratory" metaphor in the case studies that follow.

III

Pärt's earliest applications of twelve-tone technique took place in the early 1960s, when dodecaphonic music was still officially regarded by Soviet authorities as the epitome of undesirable "formalism."[29] It is not overly surprising, therefore, to find him using the less surveilled realm of film music in these

Figure 1. Transcription from the archival manuscript of *Väike motoroller* (ETMM M238:2/47).
Used with the permission of the Estonian Theatre and Music Museum.

years as a forum for serial experiment. The opening pitches of the first cue
in Pärt's very first film score spell out a tone row.[30] Still more interesting is
his second commission, also from 1962, written for a short children's puppet
animation entitled *Väike motoroller* (*The Little Motor-Scooter*).[31] As Kristina
Kõrver writes, this film depicts the adventures of a brave little scooter, which
falls out of a truck's cargo of toys and has to go searching for children by itself.[32]
Pärt's energetic score, written for chamber ensemble and child's voice, consists
mainly of the simple diatonic gestures one might expect, along with a catch-
ily strophic "travel song" that links the scooter's various adventures. However,
Väike motoroller also contains twelve-tone material, most notably beneath a
fight scene in which the scooter intervenes to rescue a terrified young girl being attacked by a snake.

An excerpt from my transcription of this passage from the archival manuscript appears in Figure 1. It is based on the exact tone row used by Pärt for
both his major concert works from the following year — *Perpetuum mobile* and
Symphony No. 1. Not only is the row identical, but certain techniques are also
shared. For instance, both the film score and the symphony contain points of
strict canonic imitation using the prime form of the row. This excerpt, I suggest, already fits the "creative laboratory" idea: It shows Pärt doing preliminary

Figure 1. (continued)

Figure 1. (*continued*)

exercises with a specific tone row before producing full-scale concert works based on the same material. This inaugurates what will become a pattern: Recall that a similar gestation period separates the music of *Värvilised unenäod* and *Sarah Was Ninety Years Old*. At a minimum, then, this early commission establishes the connectedness of Pärt's film and concert work in the context of dodecaphony. Just as significantly, the character of Pärt's twelve-tone technique was to change dramatically between *Nekrolog* (1960) and *Perpetuum mobile* (1963). In the earlier work, tone rows are superimposed on a fairly conventional symphonic dramaturgy in a manner described by Schmelz as "rather superficial."[33] *Perpetuum mobile*, by contrast, is utterly rigorous — the most tinged by multiple serialism of all Pärt's compositions. Anyone inclined to map this technical journey will find no help in Pärt's list of concert works, which contains virtually nothing for 1961 and 1962. Only "unofficial" scores such as *Väike motoroller* can yield, in their intertextual relationships, some insight into the composer's activities over these more extended periods of time.

In programmatic terms, *Väike motoroller*'s twelve-tone passage would seem to be politically innocuous. The use of a jagged, dissonant style for the relevant scene is both affectively appropriate and easily comprehended. In 1962, however, Pärt had been publicly rebuked by Soviet music's highest official, Tikhon Khrennikov, for using tone rows in *Nekrolog*.[34] Among other things, Khrennikov's speech argued that not even an ideologically "correct" program — such as *Nekrolog*'s possible depiction of antifascist struggle — could ever justify the use of serialism. The prescriptions of socialist realism in music were of course notoriously vague, and Khrennikov's simultaneous praise for the oratorio *Maailma samm* ironized (or plainly contradicted) his own position. In this light, *Väike motoroller* is an important document of Pärt's response to heightened and confusing scrutiny. It shows him, like others of his generation, refusing to abandon serial exploration while also giving himself a valuable protection: the comparatively unpatrolled realm of film scoring. The clear opposition of "good" diatonicity and "bad" dodecaphony in *Väike motoroller* entirely lacks the layer of metacommentary that Pärt would venture some years later in *Credo* (in which these two "opposites" converge ambiguously through a tone row built from the cycle of fifths). It seems to me that Pärt's use of an illicit technique in this early film score had nothing to do with political position taking. Rather, he was negotiating multiple institutional pressures in such a way as to exist both "within" and "beyond" authoritative discourse — maintaining his personal interest in self-imposed compositional strictures, articulated in this period through dodecaphonic means, by tethering them instead to projects that would be less sensitively registered than his previous orchestral showpieces. This is the kind of corrective to "binary socialism" for which Yurchak, Hakobian, and Schmelz all argue in their readings of Soviet cultural production. And this account, I suggest, is also capable of feeding back into broader understandings of how Pärt's activities were enabled and constrained in the 1960s and, ultimately, how his twelve-tone music laid the groundwork for the tintinnabuli style.

In this connection, the coexistence of dodecaphony and diatonic children's music in *Väike motoroller* is deeply significant. Since most of it is unpublished, one easily forgets the vast amount of children's music Pärt wrote (and improvised) in the Soviet Union, both for film and stage.[35] Yet the composer himself does acknowledge and value parts of this oeuvre. The socialist realist cantata *Meie aed* (*Our Garden*) was never withdrawn.[36] Talking to Peter Quinn, it was children's music that Pärt singled out when reminiscing about "beautiful moments" in his film scores. In 2015, furthermore, the choirmaster Kadri Hunt persuaded him to release several of his Soviet-era songs for children (including the *Väike motoroller* travel song) in fresh arrangements.[37] All of this means, I suspect, that the balance of extant critical literature underestimates Pärt's expertise in writing for children, refined in the 1960s, as a formative and enduring component of his compositional identity. Indeed, it is precisely because it sets children's music alongside twelve-tone experiment that *Väike motoroller* foreshadows tintinnabuli's main aesthetic qualities more comprehensively than any of Pärt's formal concert works from the early 1960s, save perhaps *Solfeggio*. Within this single film score, the diatonicism, elemental transparency, and idiomatic simplicity of Pärt's music for children complement the deterministic pitch structures, rationalized processes, and strict designs of his dodecaphony. The archetypal Pärt narrative, with its rupture-and-rebirth account of tintinnabuli, overlooks the strength of such links, instead preferring to derive tintinnabuli aesthetics primarily from the spiritual and early music encounters of Pärt's so-called self-discovery period. *Väike motoroller* urges an alternative account, in which the play of influences is far more complex, embedding tintinnabuli's origins firmly within a Soviet culture in which a composer could be a committed avant-gardist one day and write children's songs the next.

Not long after *Väike motoroller*, Pärt scored another short puppet animation entitled *Just nii!*, or "Just So!"[38] This film is a satirical critique of Soviet bureaucracy, and, although many of its cultural cues aim specifically at Estonian audiences, it is unmistakably ruthless in portraying brainless sycophancy and the triumph of the mediocre. Pärt heightens this critical bite through an acid, almost vicious score that he aptly described as "dodecaphonic jazz."[39] Attempting such a combination of bourgeois forms was certainly audacious in 1963, and the music, scored for saxophone trio, trumpet, trombone, vibraphone, drum set, piano, and bass, would have stood no chance of acceptance for concert performance. Yet Pärt clearly found the project meaningful, choosing to show *Just nii!* to the visiting Luigi Nono (the subsequent dedicatee of *Perpetuum mobile*).[40] With this commission, then, we find Pärt once more in his creative laboratory, pursuing the serial experimentation by which he was then so preoccupied.

Three aspects of the music are especially striking, the first of which is Pärt's effective use of genre as a tool of commentary. The jazz stylings are totally incongruous with the film's office-block setting, and this instantly encourages viewers to understand what they are seeing as carnivalesque and overperformed. But because of its grounding in tightly organized twelve-tone rows, the *Just nii!* score often sounds slightly unhinged even from jazz conventions, creating a still more grotesque overall effect: Figure 2, an excerpt from the opening cue transcribed from the archival score, is illustrative. Much more could be made of this coincidence of extreme order and extreme absurdity, which has been identified by Hakobian, among others, as a recurrent, "gnosiological" theme of Soviet art (certainly, it has little to do with crude conformity-versus-resistance axes of the type critiqued by Yurchak). A second notable feature of *Just nii!* is Pärt's decision to ground even the most conventional gesture-painting devices in strict twelve-tone material. This contrasts with the earlier *Väike motoroller*, whose serial content was limited to one specific affective requirement. Pärt thus increases the degree of self-discipline from one film score to the next, challenging himself in *Just nii!* to generate all his moods and effects from a single row. That trajectory is quite consistent, I suggest, with the differences in serial technique perceptible between *Nekrolog* and *Perpetuum mobile*. Accordingly, *Just nii!*'s snapshot of craftsmanship under development once more anticipates Pärt's drive toward restraint and, ultimately, his tintinnabuli aesthetics.

Figure 2. Transcription from the archival score of *Just nii!* (ETMM M238:2/35). Used with the permission of the Estonian Theatre and Music Museum.

The third point of significance concerns language. *Just Nii!* is virtually wordless, but it does imply verbal communication between characters through the frequent use of streams of instrumental sounds in substitution for actual dialogue. Figure 3 shows an excerpt from one especially evocative scene in which attendees of a meeting bray with laughter: obsequious yes men colluding to ostracize a dissentient. Pärt's rationalized twelve-tone designs are fully on display here, with the R_3 row form structuring a five-part ostinato in which even the terminal pitches of the brass glissandi are accounted for. Such passages in Pärt's early film work give the lie to any notion that the coincidence

Figure 2. (*continued*)

of rigid rationality with precise affect in his music somehow emerged in 1976 from introspective silence. But this blurring of diegetic boundaries also signals something more profound. During the implied conversations, Pärt's music occludes the literality of what is being said, replacing it with tokens of generic babble. A similar concept, termed "performative shift," underpins Yurchak's account of what happened to "authoritative discourse" in the post-Stalin Soviet Union. Yurchak describes a progression toward increasingly set and ubiquitous official phraseology, in which the correct performance of cultural ritual mattered far more than the literal meanings of what was said and done. Regrettably, the niceties of this argument lie beyond this chapter's scope.[41] However, the denial of literal language to the characters of *Just nii!* in my view engages the "performative shift" phenomenon, thus accentuating the film's satire. Both in its analytical detail and in its effect on the viewer, therefore, Pärt's score appears absolutely central to the film's status as a text that both documents and comments upon some very remarkable cultural conditions.

Taken together, *Väike motoroller* and *Just nii!* bear out in some detail the hypothesis of creative flow between Pärt's different spheres of compositional activity. In both scores Pärt hones various skills — some particular to film music, others reverberating through his concert music all the way to the tintinnabuli

Figure 3. Transcription from the archival score of *Just nii!* (ETMM M238:2/35). Used with the permission of the Estonian Theatre and Music Museum.

style. Each score also prefigures specific innovations in Pärt's concert music by about a year: in *Just nii!*'s case, the combination of row-based designs with various stylistic pastiches (think of the 1964 *Collage über B-A-C-H*). In short, the "creative laboratory" metaphor seems thoroughly vindicated by these early film commissions. I next turn to examples from a more self-evidently crucial period in the history of tintinnabuli—the late 1970s.

IV

The last feature film score Pärt wrote before emigrating was for *Navigaator Pirx*. The polyglot titles of this film, a coproduction involving several different studios and countries, seem to render best in English as "Pilot Pirx's Inquest." Released in 1979, its plot derives from a 1966 short story by the Polish science-fiction writer Stanislaw Lem.[42] Briefly put, it situates humanity on the threshold of mass-producing intelligent androids ("nonlinears"), and its concerns and anxieties are similar to those explored in such Western films as *Blade Runner* and *2001: A Space Odyssey*. Pärt's orchestral score is for the most part unadventurous. However, there is one striking feature: another instance — adding to those found in *Väike motoroller* and *Värvilised unenäod* — of musical material shared with a concert work. This time, the piece in question is *Wenn Bach Bienen gezüchtet hätte . . .*, one of the earliest tintinnabuli pieces. The music in Figure 4, played by a harpsichord (*cembalo*) over the opening titles of *Pirx*, is clearly derived from motivic material played by the same instrument in part 1 of the 1976 version of *Wenn Bach* (and still found in Pärt's most recent revision from 2001).[43] In *Pirx*, Pärt prolongs the passage with a series of transpositions. A similar cue also arrives halfway through the film, accompanying the march of Pirx and his crew onto the launchpad of their spacecraft.

Since *Wenn Bach* preceded *Pirx*, this link by itself might well be classified as a counterexample to the "creative laboratory" model of Pärt's film music. It seems that Pärt did no more than recycle an idea from a completed concert score, and indeed, this practice would come to dominate his film output in the ensuing years. Various sources, however, seem to place *Pirx* within a rather more complex and intriguing intertextual network. In 1978, Pärt's tintinnabuli programs began to include a piece mysteriously entitled *Test*. One of *Pirx*'s titles was *Test pilota Pirxa*. Minutes of the Estonian Composers' Union meeting on May 23, 1978, offer a clear connection: Here we find Pärt's wife, Nora, explaining to attendees that *Test* was simply another title for the *Missa syllabica*, a four-part tintinnabuli setting of the Mass ordinary.[44] Musical excerpts from the Latin mass, she said, had been envisaged for certain scenes in *Pirx*, and Pärt had decided to compose this material out into an independent work. Nora Pärt went on to note that *Summa* and *Cantate Domino*, two other sacred vocal tintinnabuli works presented that day alongside *Test*, had likewise developed from remnants of the *Pirx* score. The film does indeed include religious allusions, and it is altogether tempting to conclude from this record that the *Pirx* commission helped focus Pärt's earliest adaptations of the tintinnabuli style to sacred vocal composition.[45] If so, *Pirx* was a laboratory experiment that helped fundamentally define Pärt's next two decades, at least, of music.

Figure 4. Archival manuscript of *Navigaator Pirx* (ETMM M238:2/50). Used with the permission of the Estonian Theatre and Music Museum.

Here as elsewhere, however, the Soviet-era sources are not necessarily to be accepted at face value. Other evidence around *Pirx* does not support the account in the minutes. For one thing, neither the final cut nor the soundtrack recordings contain any music related to the three sacred tintinnabuli pieces. For another, the extant written materials strongly indicate that *Missa syllabica*

comfortably predated *Pirx*. Pärt's "compositional diaries" show him completing the mass between February 12 and February 18, 1977. More generally, the
diaries reveal that he devoted much of that year to the exercise of setting sacred texts according to generative syllabic principles.[46] On February 15, 1978,
when Pärt signed off on his report of activities for 1977, he listed all three of
Test, Cantate Domino, and *Summa* as completed works.[47] Meanwhile, the extensive working notebooks for *Pirx* held at ETMM are dated between August
17 and October 12, 1978. Of these, Karnes comments that in August Pärt "was
clearly starting from a very early stage in the project."[48] All of this raises the
suspicion that the *Pirx* commission arrived in early 1978 and, with its alleged
requirements around a Latin mass, offered Pärt some helpful institutional
leverage to present items of sacred music that had been written, and even unofficially performed, the previous year.[49] This is a conjecture, though I find it
a persuasive one. In any event, the case of *Pirx* offers two important lessons for
scholars in this area. First, those making claims about the aesthetic origins of
tintinnabuli can ill afford to be too credulous in gauging Pärt's account of the
tricky professional circumstances that faced him in the late 1970s. At certain
times it has suited him to associate his concert music closely with *Pirx*; at others he has disavowed the film completely. Second, the notion of film scoring
as "creative laboratory" need not operate solely at the level of notes in scores.
In this instance, the *Pirx* commission may well have helped facilitate more
public rehearsals, performances, and criticism of Pärt's concert music — events
that might plausibly, and quite independently of any historical remnants, have
contributed to the tintinnabuli style's further development.

While answers to these and other questions surrounding *Navigaator Pirx*
seem destined to remain elusive, the music for *Jäljed lumel* (*Footsteps in the
Snow*) offers indisputable evidence that Pärt did use film work as a proving
ground for tintinnabuli material in this period. A short documentary from
1978, *Jäljed lumel* is centered on a renowned Estonian actor, Ants Jõgi, who
had recently turned eighty-five and moved into an old-age home. The director, Leida Laius, struck a reflective tone, to which Pärt responded with a restrained and economical score written mainly for cello and piano. The cues
are unhurried and even, consisting mainly of long *sostenuto* lines and a regular
arpeggiated accompaniment. There is no archival manuscript, but transcription soon reveals that the score — uniquely, I suspect, within Pärt's Soviet film
output — submits to conventional tintinnabuli analysis. As Figure 5 illustrates,
there are well defined M- and T-voice relationships at work throughout *Jäljed
lumel*'s music.[50] While a full description of the design is not feasible here, it is
worth noting for present purposes that both cello and piano work in units of
six pitches with determinable interrelationships. Within a given six-note piano

Figure 5. Transcription from the score of *Jäljed lumel* (EFA 2980). Used with the permission of the Estonian Film Archive, National Archives of Estonia.

unit, for instance, pitch 1 supplies a T-voice (second position, alternating) for the cello. Pitches 3 and 5 are in diatonic sixths above the cello, while pitches 2, 4, and 6 supply T-voices for pitches 3 and 5, forming an overall arc whose shape relates back to the position of pitch 1.

The soundscape in this cue is highly characteristic of Pärt's instrumental tintinnabuli music from 1977 and 1978. *Spiegel im Spiegel* is the clearest parallel: The common structural details here include the solo/accompaniment texture; the timbral, dynamic, and rhythmic regularity; the comparatively rare use within tintinnabuli of a major key signature; the diatonic sixths; the central pitch to which the solo instrument returns at fixed points; the melodic symmetry around that central pitch; and the construction of the piano line as a hybrid of M-voice and T-voice notes (or, to borrow Anabel Maler's succinct phrase, "compound tintinnabulation").[51] *Jäljed lumel* is bound to other nearby tintinnabuli works too: In the 1977 *Arinushka* variations, for example, we find shared features including the symmetrical shift from minor to major and the repeated anapestic rhythmic pattern. The mutual permeability of film and concert work is once again obvious. Finally, the line played by the cello in *Jäljed lumel* is something of a smoking gun. Figure 6 presents the cello notes across the full cue from which Figure 5 is drawn. As this schema clarifies, the six-note units have fixed internal relationships, but they also combine to form longer cycles whose overall patterning follows higher-level rules. This, too, is a compound tintinnabuli structure.[52] As with many of Pärt's melodic lines from the late 1970s, it is an open-ended and symmetrical design, fully determinable once the rules of construction are known. Most importantly of all, it is identical to the foundational pitch structure of a later tintinnabuli work — the infrequently analyzed organ mass *Annum per annum*, composed by Pärt shortly after his emigration. In *Annum per annum* the relevant material is played in the left hand and governs all of the other lines (just as the cello notes in *Jäljed lumel* ultimately determine the piano part). Each successive movement then displays the incremental outward movement from the central pitch that is also found in the film score's cycles (see Figure 7).[53] *Annum per annum* even retains the film score's shift from minor to major pitch collections, albeit in a reworking more reminiscent of *Arinushka*. This degree of structural isomorphism reveals beyond doubt that while working on *Jäljed lumel*, as with most of the films mentioned in this chapter, Pärt generated musical material that reappeared a year or two later in his concert scores.

In view of so many such examples, it is perhaps worth recalling the stolid skepticism with which Pärt has regarded attempts to study his film music. Communicating with me via the APC, for instance, he completely rejected any suggestion that the shared music from *Wenn Bach* and *Navigaator Pirx*

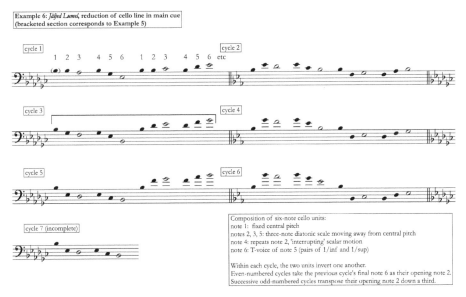

Figure 6. Reduction of the cello line in *Jäljed lumel* (EFA 2980). Used with the permission of the Estonian Film Archive, National Archives of Estonia.

Figure 7. *Annum per annum*, opening of the left-hand part in the "Gloria" movement and reduction of the left-hand part in the "Gloria" movement.

might signify any "inner or deeper" link, additionally commenting that the 1970s were "poor times" for the creation of musical material (and for that reason he often repeated ideas between his concert and film scores).[54] Those remarks, of course, are significant and must certainly bear upon the critical task of assembling and interpreting the history of Pärt's style. Yet the need to test such claims is no less certain, and *Jäljed lumel* is perhaps the clearest indicator that the tintinnabuli narrative will stand incomplete for as long as it excludes the film output. The score demonstrably documents tintinnabuli's transition from the soundworld and relatively simple design of *Spiegel im Spiegel* to the far more adventurous architecture of *Annum per annum*. A seemingly successful experiment, it therefore helped pave the way for the intricate compound tintinnabuli designs found in major 1980s works such as *Te Deum* and *Miserere*. *Jäljed lumel* straddles both Pärt's consolidation of tintinnabuli in its canonical (instrumental) form and his early attempts to reimagine and elaborate its basic principles of note against note, scale against triad, M against T. It is, in other words, precisely the kind of "creative laboratory" we might expect to find in the film output of a Soviet composer busily engaged in mapping a personal style.

With *Jäljed lumel*, this short survey reaches its endpoint. I suggest that an important pattern has emerged. Time and again, Pärt's Soviet film scores reveal something of the broader stylistic trajectories informing nearby concert works. Indeed, it seems to me apparent that the film scores are inextricably threaded through Pärt's concert output, including both the earliest premonitions and the technical crucible of tintinnabuli. The extent of the relationship undoubtedly exceeds what this chapter has presented. A challenge for future researchers in this area will be to mediate between two problematic approaches: treating Pärt's film scores as though they were equivalent to his finished concert works, on the one hand, and perpetuating their conventional status as separate and peripheral, on the other. The "creative laboratory" metaphor will, I think, continue to be a helpful guide. Naturally, there will be many for whom tintinnabuli, not film music, remains the primary reason for taking an interest in Pärt. For those readers, I hope to have shown at least that case studies from *Väike motoroller* through to *Jäljed lumel* offer a fresh perspective on two perennial questions about this famous style: How did it arise, and from where? Pärt's Soviet film scores are rich sources, and we learn much by restoring them to their deserved place in our narratives.

V

It is not widely known that Pärt's film scoring career continued after emigration. Since 1980, he has composed music for at least two further films: the American-made *Rachel River* (1987, directed by Sandy Smolan) and the British-made *A Kind of Hush* (1999, directed by Brian Stirner).[55] The first of these, *Rachel River*, has been described as "the richly emotional story of a suddenly single young woman contending with a pair of awkward, would-be suitors" that is "played out against the harsh realities and simple pleasures of an isolated northern Minnesota village."[56] *A Kind of Hush* is far grittier, following the vigilantism of "six teens who were traumatized by child abuse early in their lives."[57] In both films, Pärt's score heavily cannibalizes his own earlier concert works — *Fratres*, *Tabula rasa*, the *Arinushka* variations, and *Cantus* in

Rachel River; Psalom, Cantus, Mein Weg . . ., and *Perpetuum mobile* in *A Kind of Hush*.[58] (*A Kind of Hush* also features the heavier sounds of the alternative rock band Radiator, as well as the rarity of Pärt playing an electronic keyboard under Jan Garbarek's saxophone.)[59] However, both films, especially *Rachel River*, also contain newly written music, much of which is clearly within the tintinnabuli soundworld.[60] My impression of these passages is that they combine classical Pärtian forms (proportional canons, widening scalar arcs) with an unusually rich instrumentation that seems to anticipate major works of the late 1980s and early 1990s such as *Miserere* and in particular *Litany*. Given the balance of original and recycled material across these two films, it is clear enough that they were not major sites of technical experiment. Next to the Soviet commissions, their significance is surely modest. Nevertheless, *Rachel River* and *A Kind of Hush* are worth mentioning, suggesting as they do that the act of emigration may not have put a decisive end to Pärt's long-standing habit of using film work as a "creative laboratory." It seems to me quite possible that a fuller examination of his small post-1980 output of original film music would be of interest to this chapter's wider arguments.

The idea that quitting his Soviet environment did not terminate the relationship between film scoring and Pärt's concert music but perhaps instead simply altered its character is an important one and a suitable point of closure for this chapter. A major focus of Pärt reception discourse has been the frequent use made of tintinnabuli compositions in cinema soundtracks. Evidently, much of this music works well in films even when it was not specifically written for them. The most comprehensive study here is Maimets's, which convincingly links aspects of tintinnabuli with a variety of metaphysical and ontological topics regularly explored by directors and encapsulated by her phrase "sphere of beyond."[61] I believe that an increased knowledge of Pärt's early career as a regular film composer might well aid scholars who wish to consider why tintinnabuli has proven so attractive to so many filmmakers. The creative environment from which tintinnabuli emerged included any number of cinematic soundscapes, and the historical evidence demonstrates that Pärt honed his film-scoring craft at one and the same time as he explored materials relating to his concert works. It is tempting to suggest, then, that tintinnabuli has always been innately "cinematic" in some way and that the many film afterlives of tintinnabuli works, realized through directorial reuses that themselves are all acts of reception, simply provide confirmation of this. If that is so, then the legacy of Pärt's Soviet film scores persisted long after 1980. It is by venturing such a critical step, I think, that the present study might be brought into productive dialogue with previous scholarship on tintinnabuli and its meanings, rather than existing solely as counternarrative.

Notes

1. Paul Hillier, *Arvo Pärt* (Oxford: Oxford University Press, 1997).

2. Pärt's list of works is maintained by the Arvo Pärt Centre at https://www.arvo part.ee/en/arvo-part/works/. In 2019 this page was updated to include Pärt's film scores and other previously unlisted music. Research for this chapter was carried out in 2015 and 2016.

3. Hillier, *Arvo Pärt*, 73–74, 29.

4. See, for example, Peter C. Bouteneff, *Arvo Pärt: Out of Silence* (Yonkers, NY: SVS, 2015), 86–87.

5. Kaire Maimets, "Tasakaal su ümber ja su sees: *Ukuaru*," *Teater. Muusika. Kino* (June 2004): 92–100 (continued in July 2004, 77–85). Maimets also wrote her graduate dissertation on *Ukuaru*.

6. I am very grateful indeed to Doug Maskew for his help with obtaining some of these, along with many other useful audiovisual sources.

7. See, for example, Kaire Maimets-Volt, *Mediating the "Idea of One": Arvo Pärt's Pre-Existing Music in Film*, dissertation 4 (Tallinn: Estonian Academy of Music and Theater, 2009), 10.

8. See Immo Mihkelson, "A Narrow Path to the Truth: Arvo Pärt and the 1960s and 1970s in Soviet Estonia," trans. Triin Vallaste, in *The Cambridge Companion to Arvo Pärt*, ed. Andrew Shenton (New York: Cambridge University Press, 2012), 20–21.

9. It seems to me that the writings of Hillier, Brauneiss, and Bouteneff may all to some degree accept critical silence on certain topics as the price of closer access to the composer.

10. Andrew Shenton, "Arvo Pärt: In His Own Words," in *The Cambridge Companion to Arvo Pärt*, ed. Andrew Shenton (New York: Cambridge University Press, 2012), 111.

11. See Stephen Wright, "Stylistic Development in the Works of Arvo Pärt, 1958–1985," MA thesis, University of Western Ontario, 1992, 53–55; Enzo Restagno, "Arvo Pärt in Conversation," in *Arvo Pärt in Conversation*, ed. Enzo Restagno et al., trans. Robert Crow (Champaign: Dalkey Archive, 2012), 23; Peter Quinn, "Arvo Pärt: The Making of a Style," PhD diss., University of London, 2002, 11; Peter J. Schmelz, *Such Freedom, If Only Musical: Unofficial Soviet Music during the Thaw* (New York: Oxford University Press, 2009), 305–6.

12. Tatiana Egorova, *Soviet Film Music: An Historical Survey*, trans. Tatiana A. Ganf and Natalia A. Egunova (Amsterdam: Harwood Academic Publishers, 1997).

13. In Mihkelson, "A Narrow Path to the Truth," 20.

14. Alexei Yurchak, *Everything Was Forever, Until It Was No More: The Last Soviet Generation* (Oxford: Princeton University Press, 2006), 4–8. See Yurchak's chapter 1 for a fuller introduction to these issues.

15. Levon Hakobian, *Music of the Soviet Age, 1917–1987* (Stockholm: Melos Music Literature, 1998).

16. See, further, Oliver Kautny, *Arvo Pärt Zwischen Ost und West: Rezeptionsge-schichte* (Stuttgart: Metzler, 2002), 160–64.

17. Hillier, *Arvo Pärt*, 30–32.

18. On *Maailma samm*, see Christopher J. May, "System, Gesture, Rhetoric: Contexts for Rethinking Tintinnabuli in the Music of Arvo Pärt, 1960–1990," DPhil thesis, University of Oxford, 2016, 105–7, 138–43.

19. See, for example, Schmelz, *Such Freedom, If Only Musical*, 53–54.

20. Katherine Verdery, *What Was Socialism, and What Comes Next?* (Princeton, NJ: Princeton University Press, 1996), 94. Verdery is frank about the effect on scholarship of the Cold War's "cognitive organization of the world"; notably, this book is grounded in her skepticism of a widespread 1990s assumption among "transitologists" that formerly socialist countries would (or, indeed, should) inevitably become free-market capitalist democracies (16). As of 2019, her position appears well founded.

21. See, for example, Hillier, *Arvo Pärt*, 52, 74, xi.

22. Donald Macleod, "Arvo Pärt (1935–)," *Composer of the Week*, aired June 22–26, 2015, on BBC Radio 3 (five one-hour episodes, quotations drawn from episodes 2 and 3).

23. See, for example, Leopold Brauneiss, "Tintinnabuli: An Introduction," in *Arvo Pärt in Conversation*, ed. Enzo Restagno et al., trans. Robert Crow (Champaign: Dalkey Archive, 2012), 107–62.

24. Andrew Shenton, ed., *The Cambridge Companion to Arvo Pärt* (New York: Cambridge University Press, 2012). Immo Mihkelson's chapter is a notable exception.

25. Not *Für Alina*, as is widely claimed. For more on *Sarah Was Ninety Years Old*, which at that time bore the title *Modus*, see C. J. May, "Sonic Embodiment: *Sarah Was Ninety Years Old*," in *Arvo Pärt's White Light: Media, Culture, Politics*, ed. Laura Dolp (Cambridge: Cambridge University Press, 2017), 179–211.

26. Kevin C. Karnes, *Arvo Pärt's Tabula Rasa* (New York: Oxford University Press, 2017), 82, 125.

27. Kristina Kõrver, email to the author, November 27, 2015. Compare Egorova, *Soviet Film Music*, xii.

28. See Wright, "Stylistic Development," 55.

29. See Schmelz, *Such Freedom, If Only Musical*, chap. 2 ("The Dam Bursts: The First and Second Conservatories").

30. The film was Leida Laius's *Õhtust hommikuni* (*From Evening to Morning*). See, further, May, "System, Gesture, Rhetoric," 146–48.

31. Available at https://www.youtube.com/watch?v=9eJRPQGTOw4.

32. Kristina Kõrver, liner notes to Arvo Pärt, *Lapsepõlve lood* (*Songs from Childhood*) (Eesti Rahvusringhääling, Arvo Pärdi Keskus, 2015, CD). This release, which commemorates Pärt's music for children, was an unexpected and welcome addition to his discography.

33. Schmelz, *Such Freedom, If Only Musical*, 132.

34. The denunciation was made in Khrennikov's address to the Third All-Union Congress of Soviet Composers on March 26, 1962. Tikhon Khrennikov, "On the Way to the Musical Culture of Communism," *Current Digest of the Soviet Press* 14 (1962): 14–17.

35. See Toomas Siitan, "Laudatio," given on January 19, 2011. http://www.arvopart.ee/en/arvo-part-2/selected-texts/laudatio-by-toomas-siitan/.

36. A good many Soviet creative figures found sanctuary in the natural overlap between socialist realism and children's art.

37. Released as Pärt, *Lapsepõlve lood*. See also Quinn, "The Making of a Style," 11.

38. Other possible renderings of the idiomatic title in English are "Quite so!" or "That's right!"

39. Restagno, "Arvo Pärt in Conversation," 23.

40. See also Mihkelson, "A Narrow Path to the Truth," 20–21.

41. See further Yurchak, *Everything Was Forever*, chap. 1.

42. See Stanislaw Lem, "Inquest," in *More Tales of Pirx the Pilot*, trans. Louis Iribarne, Magdalena Majcherczyk, and Michael Kandel (London: Secker & Warburg, 1983), 90–161.

43. For a detailed study of *Wenn Bach* in its various versions, see May, "System, Gesture, Rhetoric," chap. 1.

44. My study of these sources is independent of (and differs from) the account given in Kautny, *Arvo Pärt Zwischen Ost und West*, 123–25. The minutes are held at ETMM.

45. Excluding 1976's *Calix*, where the dies irae text was concealed under solmization syllables.

46. "Compositional diaries" is the APC's phrase for the working notebooks kept by Pärt. I am grateful to Kevin Karnes for sharing with me his detailed research into Pärt's compositional activities and performances in 1976 and 1977. See further his chapter in this volume.

47. This report is held at the Estonian Composers' Union.

48. Kevin Karnes, email to the author, November 2, 2017.

49. The composition of sacred music remained problematic in Soviet Estonia in the late 1970s, as witness Pärt's habitual use of neutral titles in these years. The May 23, 1978, minutes cited previously go on to record an uncomfortable discussion in the wake of *Test*'s "official" presentation. As Karnes has established, all three tintinnabuli works linked to *Pirx* were performed in October 1977 as part of an avant-garde music festival in Riga, Latvia.

50. Within Hillier's standard analytical description of tintinnabuli syntax, an M-voice is a scalar line moving from or to a central pitch, while a T-voice is a line confined to the pitches of a given triad and derived from its related M-voice by fixed principles.

51. Anabel Maler, "Compound Tintinnabulation in the Music of Arvo Pärt," MA

thesis, McGill University, 2012. Pärt's "compositional diaries" contain a twelve-page sketch for *Jäljed lumel*, dated April 12, 1978, confirming its chronological proximity to *Spiegel im Spiegel* (APC Notebook 31, April 8–29, 1978).

52. The notes of the cello function as an M-voice in relation to the piano but as both M-voice and T-voice notes in relation to each other.

53. On the analysis of *Annum per annum*, see Maler, "Compound Tintinnabulation"; and Marguerite Bostonia, "Musical, Cultural, and Performance Structures in the Organ Works of Arvo Pärt," DMA diss., West Virginia University, 2009.

54. Kristina Kõrver, email to the author, November 27, 2015. At the time of this exchange I had not yet traced the *Jäljed lumel* links.

55. See the IMDb entries at http://www.imdb.com/title/tt0093816/ (*Rachel River*) and http://www.imdb.com/title/tt0178675/ (*A Kind of Hush*).

56. Phrasing from the synopsis that accompanies the commercial disc of *Rachel River*.

57. Phrasing from the film's IMDb page.

58. Material from *Psalom* in particular was recorded specifically for use in the film, as surviving recording takes demonstrate.

59. Listed in the credits as *Untitled for Saxophone and Keyboard*. Maskew notes that the same music is credited as *Aetos* in Dorian Supin's documentary *Who Is Arvo Pärt?*

60. An example of the nontintinnabuli material is heard in the trailer for *Rachel River*: https://www.youtube.com/watch?v=-c7aIUEdx7E.

61. See Maimets-Volt, *Mediating the "Idea of One,"* 141ff.

5

Arvo Pärt's Tintinnabuli and the 1970s Soviet Underground

Kevin C. Karnes

The Tallinn premiere of Pärt's "tintinnabuli" music on October 27, 1976, was an auspicious event in the life of the city. Staged in the Estonia Concert Hall, the capital's most prestigious venue, and featuring the Tallinn Chamber Choir, it was performed by Hortus Musicus, Andres Mustonen's celebrated early-music ensemble, which had enjoyed the sponsorship of the Estonian Philharmonic since its founding in 1972. To borrow from the musicologist Peter J. Schmelz, the premiere, sanctioned and supported by the Ministry of Culture, was as "official" an event in the Estonian SSR as any concert could be.[1] In many respects, the official status of the Tallinn premiere was unsurprising. After all, Pärt was an award-winning composer whose work had been alternately celebrated and censured by Soviet authorities for over a decade. Still, despite the fact that Pärt had not had a concert premiere in over three years, some had doubts about Pärt's latest stylistic turn. "The concert on 27th October did not cause a sensation," the Estonian journalist Immo Mihkelson remembers. As Mustonen later recalled of the event, "it was not yet clear if anything would come of this."[2]

As Mihkelson notes, the Tallinn premiere was previewed two days earlier, when Hortus Musicus presented a shorter suite of Pärt's tintinnabuli-style works at Tartu University. The largely academic audience at the university anticipated the distinctive character of the audience in Tallinn, which Mihkelson recalls as "noticeably younger than the typical concert-goer."[3] The Estonian musicologist Toomas Siitan likewise describes a distinctive cast to the crowd attending many of Pärt's early tintinnabuli performances. It consisted largely of individuals more likely to be seen at concerts of progressive rock than at the philharmonic's classical programs, "young people com[ing] from throughout

Estonia" to hear "something real," he remembers, "something fresh."[4] Even the Tartu University concert was not the first public performance of Pärt's tintinnabuli-style music, however. As the composer recorded in his musical diaries, *Sarah Was Ninety Years Old* — called *Modus* on the programs in Tartu and Tallinn — had its premiere, *unofficially*, on April 27, 1976, at a festival of new music held at the Anglican Church in Riga's Old Town.[5] At that time, the building was not functioning as a church at all but as home to the Student Club of the Riga Polytechnic Institute. An architecture student named Hardijs Lediņš had been holding a wildly popular series of discotheques there since the 1974–1975 academic year. In October 1977, the Student Club would host a second discotheque new-music festival, which would feature a slate of further tintinnabuli premieres: *Arbos*, *Cantate Domino canticum novum*, *Fratres*, and *Summa* were first performed at the festival, along with a mysterious tintinnabuli work called *Test*. As the pianist Alexei Lubimov, a festival participant, recalls of this and other experiences he had at the Riga Student Club, "it was as if we were discovering a whole new universe."[6]

In this chapter, I will focus on these and other early, largely forgotten engagements between Pärt, tintinnabuli, and student culture in the 1970s USSR. I will document the embrace of Pärt's new style among students and young artists, and I will suggest that an informal network of Soviet youth and experimenting musicians played a crucial role in fostering and promoting his tintinnabuli works between the time of the official premieres of October 1976 and the international success of *Tabula rasa* in late 1977. While not denying the singularity of Pärt's achievement with his new compositional language, I aim to chip away at the persistent image of Pärt as a solitary, isolated figure at the time of his greatest creative breakthrough. While his tintinnabuli experiments alienated him from some — though by no means all — of his professional colleagues, they simultaneously garnered him the embrace of a community whose sound experiments resonated with his own and whose attendant spiritual searching paralleled that which had brought him to his new style in the first place.[7] I will suggest that the Latvian capital became a kind of laboratory for the composer during a period of approximately eighteen months, home to what I will call, following the Russian painter Ilya Kabakov, an "underground" (*podpol'nyy*) space of relative creative freedom — a space in which some of Pärt's most important premieres took place unofficially, as if unnoticed by the Ministry of Culture, unpoliced by Soviet authorities.[8] And I will suggest that Pärt's Riga festival concert of October 28, 1977, was particularly important in his history because many of his works premiered that night were openly religious, even sacred in nature. Indeed, it was within the underground space of Lediņš's Riga discotheque that Pärt came out publicly, for the very first time, as a Christian composer of overtly and explicitly devotional music. The story I tell in this chapter is a story about Pärt and his new tintinnabuli style. But it is also a story about Riga and of the spaces it afforded, at least for a moment, for the kind of creative coming-out Pärt ventured: for religious utterance and aesthetic experiment — in short, for the freedom to be an artist.

The Mystery of *Test*

I'll begin with a mystery. On February 15, 1978, Pärt filed his annual report with the Union of Soviet Estonian Composers, listing all of his works composed during the previous year.[9] His list indicates nine new compositions in 1977, eight of which we recognize today. These include *Cantus in Memory of Benjamin Britten* and the double concerto *Tabula rasa*. Below those, Pärt listed *De profundis* and the *Variations for the Healing of Arinushka*. Numbers 5 through 9 on his list are interesting for the fact that they were premiered all together, in October 1977, in the Latvian capital. Piece number 5 is *Cantate Domino*, numbers 6 and 7 are *Fratres* and *Arbos*, and piece number 9 is *Summa*. Number 8, however, would appear to be a ghost — and some unknown reader of this list even penciled in a question mark next to that entry on the document. Pärt recorded the title of the eighth work on his list as *Test*, which he described as "music for a Polish-Soviet film." Three months later, on May 15, 1978, a work called *Test* appeared once again, this time as part of a suite of tintinnabuli-style compositions performed by Hortus Musicus in Tartu.[10] Ever since the official unveiling of Pärt's tintinnabuli style a year and a half before, what the composer had been calling his tintinnabuli "suite" (*oopus* in Estonian, *opus* in Russian) had comprised a varying list of short compositions. *Trivium*, *Für Alina* (*Aliinale*), and *Pari intervallo* had all been performed as part of the suite at one time or another. Now, in Tartu in May 1978, *Test* was taking a turn on the program.

A work called *Test* appears nowhere in Pärt's official catalogue of compositions. The question of its identity nagged me. Just what was this composition, called *Test*? Where did it come from, and where did it end up? We find a hint in transcripts from a meeting of the Estonian Composers' Union on May 23, 1978, eight days after the Tartu concert where the work was performed. There, as Christopher J. May has documented, the composer's wife, Nora, revealed something surprising. *Test*, she claimed, had been composed as a Latin Mass. Moreover, it had been intended for use as part of one of Pärt's film soundtracks — specifically, a jointly produced Polish-Estonian science-fiction film called *The Test of Pilot Pirx*.[11] So why, then, would Pärt have elected to compose a mass for a film? The movie in question, *Pilot Pirx*, released in

1979, follows the journey into space of an astronaut named Pirx and his crew, which consists of a mix of humans and potentially untrustworthy androids. Pirx struggles to tell the two groups apart. To be sure, the movie includes a fair amount of talk about religion, as Pirx identifies faith in God as a distinctly human attribute, and he questions his crew members about their religious convictions as he tries to distinguish man from machine. Several hundred pages of manuscript materials related to Pärt's work on the film are preserved at the Estonian Theatre and Music Museum in Tallinn, ranging from rudimentary thematic sketches to fair copies of fully orchestrated passages.[12] The earliest of these materials appears to date from August 1978. Yet a careful examination of these materials reveals not one of their pages to preserve a trace of anything resembling a mass, nor do the sketches preserve any hints of Pärt's tintinnabuli style, beyond a quotation from *If Bach Had Kept Bees* (*Wenn Bach Bienen gezüchtet hätte*) (1976) at the very start. Having thus mined the archives in Estonia, I was left facing the mystery with which I began. What was this composition called *Test*? Where did it *really* come from, and where did it end up?

The Mass at the Disco

Several months later, I traveled to Riga to visit the archives of Hardijs Lediņš (1955–2004), an artist, composer, DJ, and producer of some of the first New Wave recordings in the Soviet Union (Figure 1). Back in the winter of 1974–1975, as an architecture student, Lediņš founded a popular series of discotheques at the Student Club of the Riga Polytechnic Institute, which made its home in the city's disused Anglican church. What drew me to Lediņš was the fact that *Modus*, Pärt's very first work to be labeled by the composer with the heading "tintinnabuli" (as *Tintinnabulum 1*), had its first-ever performance, under the title of *Saara*, in April 1976, in the unofficial venue of a festival of new music organized by Lediņš under the umbrella of his disco.[13] The identities of the performers are obscure today. In a recent conversation, Lubimov recalled playing the premiere with the celebrated soprano Lydia Davydova.[14] However, Pärt remembers that the singer on the premiere was clearly pregnant, whereas Davydova, born in 1932, was not at the time.[15] A photograph preserved in Lediņš's archive at the Latvian Centre for Contemporary Art in Riga captures Pärt during the setup for one of the other 1976 festival events; Pärt is on the left-hand side (Figure 2).[16] It was also Lediņš who organized the Riga concert of October 1977 at which *Test* and four of Pärt's other new tintinnabuli works were premiered together. That concert was part of a second festival of contemporary music organized by Lediņš at the Riga Polytechnic Student Club, and Lediņš's archive preserves a complete typescript schedule of the

Figure 1. Hardijs Lediņš around 1980. Latvijas Laikmetīgas mākslas centrs/Latvian Centre for
Contemporary Art, Riga.

festival's events. Among them, on October 28, at 10:30 p.m., Hortus Musicus
was scheduled to perform a concert, with the note "A. Pärt's music [is] on the
program" ("Programmā A. Pērta skaņdarbi," in Lediņš's original).[17]

If a printed program for the concert of October 28 was ever produced, I have
not found a copy. But the private archive of a Lediņš family friend preserves a

Figure 2. Pärt (left) during setup for one of the 1976 festival events. Latvijas Laikmetīgas mākslas centrs/Latvian Centre for Contemporary Art, Riga.

reel of magnetic tape on which the concert was recorded.[18] On the outside of the box that houses the tape, in what looks like Lediņš's handwriting, the first piece listed was the *Passionslieder*, by the Moscow-based composer Vladimir Martynov, which likewise had its premiere at the festival. After that, yet another version of Pärt's tintinnabuli suite was performed by Andres Mustonen and Hortus Musicus, with the program listed on the box in a mix of handwritten Latvian, Latin, and Russian. This time, the suite consisted of *Arbos*, *Cantate Domino*, *Summa*, *Fratres*, *In spe*, and *Pari intervallo*. Then, confirming at least part of Nora Pärt's story to the Soviet Estonian Composers' Union, another work by Pärt was recorded, which Lediņš labeled "Mass in Six Parts (Test)"—in the original Latvian, "Mesa (6 daļās) (Tests)." Fortunately, the tape was still playable. Listening to a copy, I at last found an answer to the mystery that had prompted my months-long search.[19]

The work called *Test*, it turns out, was indeed a mass. But it was not just any mass. It was Pärt's *Missa syllabica*, which, as Pärt's musical diaries reveal, was the very first tintinnabuli composition he had completed using what has since become known as his "syllabic" method of setting texts — basically, generating the contour of a melodic line algorithmically, with input into the algorithm consisting in part of the syllable count of each word to be set.[20] As indicated in the diaries, Pärt discovered the syllabic method on February 12, 1977, and he used it that day to generate a melodic line for an Orthodox prayer. Immediately after that, he used the same technique to set the Latin *Gloria* syllabically. Between that date and February 15, he composed syllabic melodies to set the remaining texts of the Mass ordinary: kyrie, agnus dei, sanctus, and credo. He spent the next three days filling in secondary melodic lines and adding triadic tintinnabuli voices (T-voices) to them. By February 18, he had finished a complete Latin Mass, which he would later call *Test*, and later still, *Missa syllabica*.[21] For present purposes, what is important to note about Pärt's discovery of his syllabic method is that it arose not from his search for a means of adapting tintinnabuli principles to the setting of texts in general but specifically from his search for a way to set the biblical words of God and of the saints in his music, a project upon which he had embarked as early as November 11 or 12, 1976.[22] Pärt's project of these months, in other words, was avowedly religious. It was an explicitly Christian project, with his first priority, upon discovering the syllabic method, being to set the entirety of the Mass ordinary, the text that constitutes the very heart of Catholic liturgical practice.

In the 1970s, the Soviet Union was an officially atheistic state, but it was emphatically not a place devoid of religion. Already for decades, and especially in the 1970s, many Soviet artists, including musicians in Pärt's circle, had composed sacred music or participated in private religious study.[23] But to do so publicly, as the anthropologist Alexei Yurchak notes, was substantially outside the bounds of what constituted "normal" behavior for a Soviet citizen. "Religion was tolerated by the state," Yurchak writes, "but disconnected from state institutions (education, media, industry, public associations, army, bureaucracy, etc.)." Its practice, Yurchak continues, was "tolerated but viewed with suspicion and hostility."[24] Back in 1968, Pärt had gotten himself into considerable trouble for arranging a public performance of his *Credo*, a work for chorus and orchestra whose Latin text declared openly his Christian faith: "I believe in Jesus Christ . . ." The scandal that erupted afterward contributed to the greatest creative crisis of Pärt's life, a crisis he resolved only with his discovery of the tintinnabuli style in 1976.[25]

In light of this, it may seem remarkable that Pärt's musical diaries reveal his tintinnabuli project to have been a devotional project from its very start.

As late as September 12, 1976, six weeks before the official Tallinn premiere, Pärt was still considering a number of possible titles for his earliest tintinnabuli compositions, and several of those possibilities point to religious origins of the works. The choral work *Calix* was identified as a dies irae from the Requiem Mass, and sketches for the piece, first described by Merike Vaitmaa, reveal that the score was conceived as a setting of the Latin hymn. Ahead of the premiere, *Modus* was still being called *Saara*, alluding to the Book of Genesis. And *In spe*, published in 1984 under the title *An den Wassern zu Babel saßen wir und weinten*, was also, in the diaries of September 1976, identified as a kyrie from the Latin Mass ordinary.[26] For the official premiere in Tallinn on October 27, however, Pärt disguised the devotional nature of all of these compositions. All of the movements of his suite were identified only by their neutral, nonreligious titles. In *Calix*, the choir sing solfege syllables instead of its sacred text.[27]

In contrast to his perceived need to hide the religious nature of his tintinnabuli project in Tallinn, the story that unfolded for Pärt in Riga was substantially different. In the disco-festival space of Lediņš's Polytechnic Student Club, Pärt seems to have found an *underground* space, a space off the radar of Soviet officialdom, where he felt free to proclaim his faith in his music. Not only did his festival concert of October 1977 feature the first performance of the *Missa syllabica*, it also included two additional openly sacred tintinnabuli works, *Summa* and *Cantate Domino*. With this in mind, one might ask what it was about this festival space that made it one in which the composer seemed to have no fear. What made it a space in which he could venture to come out with works every bit as officially unacceptable as *Credo*, for which he had gotten into considerable trouble less than a decade before? If answers are to be found, I think, the figure of Lediņš might help us discover them.

Enter the DJ

When he first proposed his idea for a series of discotheques at the Riga Polytechnic Institute, Lediņš envisioned something different from what many might picture today. His discos would feature dancing, of course, but only in the second half of the evening. In the first half, which he called their "educational" (*tematiska*) part, he would play recordings of music (classical and pop, Soviet and Western, recent and centuries-old), feature performers from the Latvian SSR Philharmonic Chamber Orchestra and elsewhere, and lecture to attendees about what they heard. As Aina Bērziņa, head of the Student Club at the time, described the project in the Polytechnic's student newspaper in October 1974,

This year, the Student Club is organizing something new: the discotheque. It will be a new kind of recreational evening for students,
distinguished from typical events by virtue of the fact that a significant
portion of our attention will be on the educational part of evening.
We'll strive to acquaint ourselves with the newest jazz, estrada, and
pop music. We'll organize meetings with experts and discuss subjects
of vital interest today. In this way, the events will constitute something
new at our institution.[28]

To prepare for his discotheque programs and lectures, Lediņš mined the record and magazine collections of the Latvian State Library and Conservatory,
writing out in longhand scripts for delivery at the events.[29] His discotheque
programs were ambitious. One of the few that has survived, from November
1978, gives us a sense of the range of his concerns. The first, "educational" part
of the evening Lediņš devoted to the topic of "electronic musical instruments
and their use in rock." He traced the sounds and technologies of the contemporary West German band Tangerine Dream back to the experimental work
of Karlheinz Stockhausen and to points even deeper in music history.[30] From
the start, officials at the Polytechnic recognized Lediņš's project as a serious
educational endeavor, and they supported it enthusiastically. In December
1974, when he first put forward his proposal, the Polytechnic branch of the
Soviet Youth League, or Komsomol, in charge of promoting and controlling
student life at the institution, recommended approval of his plans without
any formal deliberation whatsoever.[31] The Komsomol reaffirmed its support
for Lediņš's discos repeatedly over the next two years.[32] As I was told in an interview with the university administrator Asja Visocka, who presided over the
Polytechnic's Student Club beginning in January 1978, Lediņš and his friends
were "very, very active" in shaping cultural life at the institution. And "all of
them," Visocka told me, "were upstanding."[33]

The upstanding nature of Lediņš's work carried him far in putting together
his festival of October 1977. To gain support of the Komsomol, he and his colleagues framed the event as a celebration of the sixtieth anniversary of the Bolshevik Revolution, as attested by a flyer produced to advertise the festival on the
Polytechnic campus (Figure 3).[34] As one of Lediņš collaborators in staging the
event, the Latvian violinist Boriss Avramecs, reports, the students declared the
goals of their project to be "1. To showcase the most recent achievements in
Soviet music" and "2. To foster the exchange of views and contacts between
young people of the brotherly republics."[35] Recently, in Vilnius, I showed a
photograph of the festival flyer to Donatas Katkus, who played first violin in

RPI STUDENTU KLUBS

21. oktobris — 30. oktobris 1977. g.

MŪSDIENU MŪZIKAS DEKĀDE

VELTĪTA

Lielās Oktobra sociālistiskās revolūcijas 60. gadadienai

Senās mūzikas ansamblis «Hortus musicus» (Tallina)
Viļņas stīgu kvartets
Sitamo instrumentu ansamblis. M. Pekarskis (Maskava)
Saksofonu kvartets. L. Mihailovs (Maskava)
Elektroniskās mūzikas studija (Maskava)
Vokālais ansamblis «Tonika» (Ļeņingrada)

Programmā padomju komponistu skaņdarbi

Biļetes RPI studentu klubā, Bibliotēkas ielā 2a

Figure 3. Flyer for the festival of October 1977, "dedicated to the 60th anniversary of the great October socialist revolution." Latvijas Laikmetīgas mākslas centrs/Latvian Centre for Contemporary Art, Riga.

the Vilnius String Quartet when they appeared at the festival. Not having seen the flyer in over forty years, Katkus registered a shock of immediate recognition. Then he burst out laughing. "We were crazy," he recalled of the event and of the actions and intentions of his friends and collaborators. "We could do whatever we wanted."[36] As another festival organizer, Alexei Lubimov, explained to the composer Valentin Silvestrov in a letter of July 12, 1977, just as preparations for the October events were underway, "The festival taking shape [in Riga] is being supported by people from the Komsomol. In particular, it is being organized under the banner of the 60th [anniversary], because if you are already [operating] within the sphere of the Ministry of Culture, the censor will leave you alone."[37] It is in such words from Katkus and Lubimov, I think, that we might find important clues to understanding just why and how the Student Club came to be so important to Pärt's tintinnabuli project in these years.

It is clear from the archives that Lediņš was highly esteemed in many corners of Riga officialdom, that his organizational work among students and musicians was officially supported, and that his activities were subjected to little or no careful oversight by institutional or governmental authorities. It is also clear that he understood quite well how to market his projects to officials so as to maintain their support and the freedom he enjoyed. It is within the relatively unmonitored space of his discotheque project that a little underground domain of creative expression seems to have been able to take root, one just big enough to accommodate such an important project as the first performance of Pärt's openly sacred tintinnabuli music. In this light, we might return to the ambiguous, early title of the *Missa syllabica*, premiered at the festival as *Test*. As Pärt's musical diaries make clear, the work was composed in February 1977, some eighteen months before he would even begin his work on the soundtrack for *The Test of Pilot Pirx*. But that film *did* focus on questions of religion, so I suspect that to associate his mass with that particular soundtrack, once it had already had its underground premiere and he had begun to acknowledge its existence publicly, might have been a strategic move. Namely, the putative association of the mass with the film could have offered Pärt an excuse, of sorts, for the religious nature of the work, a way of avoiding a repeat of the *Credo* scandal in the event that such danger might loom again. In effect, it shows Pärt hiding yet also not hiding, openly acknowledging the existence of his mass yet taking care to lay a foundation of plausible deniability if a concerned official should ask too many questions, listen too closely, or approach too near. In contrast, in the underground space of Lediņš's Riga festival, Pärt — like Katkus and Lubimov — felt that he could do whatever he wanted. Musically, he had felt free.

The Ritual Moment

Significantly, Pärt's was not the only sacred music to be performed at the October festival. Vladimir Martynov, a young composer from Moscow, had also had a sacred premiere at the event: his *Passionslieder* (1977), composed for soprano and a Baroque ensemble of strings and harpsichord. Comprising nine movements, each of which was based on a single G-minor, four-bar melody, Martynov's work is a highly repetitive, even hypnotic setting of a Lutheran chorale text written by the German theologian Johann Mentzer at the turn of the eighteenth century. "Der am Kreuz ist meine Liebe, meine Liebe ist Jesus Christ": "He on the cross is the one whom I love, the one whom I love is Jesus Christ." Martynov's performance, too, was captured by Lediņš on reel-to-reel tape.[38] As Martynov would recall some two decades later, he first heard Pärt's tintinnabuli music at the same Riga festival where his *Passionslieder* premiered (Pärt's tintinnabuli style was not yet widely known in Moscow), and the experience struck him as something "volcanic, tectonic."[39] The festival as a whole, as he described it in hindsight, was an event in which he and Pärt "came out together, as if in a united front, to declare a new compositional truth."[40]

It is not hard to appreciate the shock Martynov experienced when he first heard Pärt's tintinnabuli-style music in Riga, for the quasi-medieval sounds of *Test* and *In spe* would likely have suggested a commonality between Pärt's aesthetic concerns and his own, especially when he heard those works shortly after the quasi-Baroque sounds of his *Passionslieder*. And their spiritual concerns intersected as well. At that time, Martynov had yet to commit fully to Orthodox Christianity. But he had already embarked upon a spiritual journey several years before. Working at a Tajik film studio at the start of the decade, he had spent his nights listening to records with friends, especially the progressive rock of King Crimson, Tangerine Dream, and Klaus Schulze. After listening, he and his companions would talk with one another about the spiritual truths — the "Great Religious Idea" (*Velikaya Religioznaya Ideya*) — that those artists' music seemed to address.[41] Soon, back in Moscow, Martynov took a job at the electronic music studio of the Skriabin Museum. There, he formed the rock band Boomerang (in Russian, *Bumerang*) with studio colleagues and musician friends, and he continued exploring connections between musical and spiritual practice. Listening one night at the museum's studio to King Crimson's 1974 album *Red*, Martynov had a vision: of music as a "sparkling stream embracing and penetrating everything," and of the task of the composer or musician as consisting, simply, in "lowering themselves into that stream."[42]

As Martynov later recounted, he had his first inklings of the ideas that led him to compose in the idiom of the *Passionslieder* when he read a pair of standard-issue music history texts, the polemical statements traded around 1600 between the composer Giovanni Artusi and a brother of the composer Claudio Monteverdi. Technically, the Italians' debate focused on Monteverdi's then-radical treatment of musical dissonance, but it hit more broadly upon the role of any composer in relation to the texts they set. Artusi held that it was the duty of composers to subject themselves to the norms and rules of inherited tradition. Monteverdi countered that composers — as freely interpreting intellects — was entitled to bend the rules however they wanted in order to express their own interpretation of a text. Martynov, who was just then starting to think about music as a "path toward religious understanding," found himself siding emphatically with Artusi.[43] He discussed the centuries-old debate with friends whenever and wherever they were willing to listen, and he seriously pondered writing his own polemical treatise to set the historical record straight. Monteverdi's music, and most of the music that came afterward, was, as Martynov described it, "artistic music" — music created to give sounding expression to the mind or ideas of the artist. In contrast, the music of earlier ages was "magical music," as he called it, music that merely sounds *through* the composer, music that sounds the voice of God. That was the kind of music Martynov wished to write.[44]

In unknowing parallel to Pärt in Tallinn, who had begun studying medieval music around 1970, Martynov immersed himself in studies of early and pre-Baroque music in Moscow. He also began to seek, in his own compositions, ways of tapping into the sounds of earlier ages — from a time, he felt, before the advent of thinking about the composer as a self-expressing artist. This meant, for Martynov, the sixteenth century or earlier, whose sounds and gestures he sought to emulate in his own compositions (he later described his project as composing with historical "simulacra").[45] Thus the deliberately archaic sounds of the *Passionslieder*. When Martynov first heard Pärt's tintinnabuli-style music — his *sacred* tintinnabuli — in Riga in October 1977, he felt as if he had seen a vision of himself reflected in a musical mirror, as if he had heard an echo of his own spiritual quest in the resonant harmony of Pärt's compositions (Figure 4). With their joint premieres at Lediņš's festival, the event had the cast of a ritual moment. It was as though the home of the Polytechnic Student Club — an erstwhile Anglican church — had become a church once again, even if only for a couple of days, the sacred rite resounding in its walls for the first time since the country's Soviet annexation thirty-three years before.

An Ending

As Avramecs recalls, mistakes were made at the October festival. One of the its organizers reportedly dropped typescript copies of Mentzer's Lutheran text

Figure 4. Vladimir Martynov and Arvo Pärt in Tallinn, 1978. Latvijas Laikmetīgas mākslas centrs/ Latvian Centre for Contemporary Art, Riga.

from the balcony of the Anglican church into the audience below, just as Martynov's *Passionslieder* was being performed. A copy of the text is said to have made its way to the KGB. Quickly, the festival organizers and their enablers were charged with engaging in "religious propaganda."[46] To date, Latvia's KGB files remain closed to most researchers, and I have been unable to find any record of repercussions in the archives of the Komsomol, the Riga Polytechnic, or the Soviet Latvian Composers' Union, which is said to have tried to intervene in defense of the students and the artists implicated.[47] However, several of the festival's organizers later described what they experienced. As Lediņš would recall in the time of glasnost, "after the repressions" that followed the festival, "I no longer involved myself with avant-garde music."[48] Referencing the imaginary "travels" that such musical engagements had once afforded him, he explained: "Because I was an architecture student, I was called in to see the rector, and that's why I ceased my travels around the world. I was given a choice: to continue with my studies, or to organize further festivals."[49] Lediņš chose the former. For the pianist Lubimov, the repercussions were harsher. "After the festival in Riga," he explains, "I was deprived of concert trips abroad for several years."[50] Harsher still was the punishment given to the head of the Student Club, Aina Bērziņa, who lost her job.[51] The decisions behind these moves seem to have been made quietly, or in spaces still inaccessible to historians. When I spoke with Visocka, Bērziņa's successor in the post, I asked her about the circumstances surrounding the shakeup after the festival. She insisted that the event was a "completely normal festival," and she explained the cessation of Lediņš's organizational work simply: "There was," she told me, "no real desire to continue."[52] The Latvian musicologist Martin Boiko shared a somewhat different recollection with me. Although he did not attend the festival himself, he remembers arriving at the Student Club for an event shortly after the scandal broke. "One day I went. There was this lady who always sat, smiling, at the entrance of *Anglikāņi* [the Anglican church]. Now, she was crying. And that was it."[53]

As the 1977–1978 academic year drew to a close at the Riga Polytechnic Institute, changes were afoot. Working diligently to complete his degree, Lediņš continued to organize discos, but he increasingly sought venues beyond the campus, and he devoted an ever-larger share of his programs to New Wave music and dancing. In Moscow, Lubimov and his friends were busy making plans for a third festival of new music, but the events of the previous October had compelled them to look outside of Latvia entirely. With Lediņš's departure from the scene, they connected with Andres Mustonen, the director of Hortus Musicus, who leveraged his esteem with Estonian officials to open the doors of Tallinn's musical establishment to an officially sponsored, blowout event planned for November 1978.[54] The tradition of the Polytechnic festivals in which Pärt's sacred tintinnabuli project took root would continue after all, but in a new locale: Tallinn, right in Pärt's backyard.

Avramecs would not be involved in organizing the Tallinn festival of 1978. For him, as for both Martynov and Lubimov, the culminating moment for the underground music scene he helped shape would remain the Riga Polytechnic festival of October 1977.[55] Looking back from the vantage of nearly thirty years after the event, Avramecs described its significance largely in terms of its impact upon the social realm, a realm both aesthetic and spiritual and perhaps even slightly, vaguely political. Recalling the festival concert at which Martynov's *Passionslieder* was premiered, he remembered:

> The concert took place late in the evening, at the end of October. And that night, on the embankment beside the Anglican Church, there was a rehearsal for an army parade going on, because November 7 was approaching [when the anniversary of the Bolshevik revolution was to be celebrated]. You could hear the grumbling of truck drivers and commands shouted in Russian. But right next to all that, it was a little oasis.[56]

At that moment, in that ritual moment, it was as though a spiritual history interrupted decades before was revealed to be continuous after all. Underground, the sacral life of the church in Riga's Old Town resumed, momentarily.

Notes

1. An earlier version of this chapter appeared in *Res Musica* 11 (2019), https://resmusica.ee/en/res-musica-11/. On "official" and "unofficial" musical works and events in the USSR, see Peter J. Schmelz, *Such Freedom, If Only Musical: Unofficial Soviet Music during the Thaw* (New York: Oxford University Press, 2009). The program for the concert is preserved at the Estonian Theatre and Music Museum in Tallinn (Eesti Teatri- ja Muusikamuuseum, hereafter ETMM), M238:1/4. Materials related to the early activities of Hortus Musicus are preserved in ETMM, MO20.

2. Immo Mihkelson, "The Cradle of Tintinnabuli — 40 Years since a Historic Concert," https//www.arvopart.ee/en/the-cradle-of-tintinnabuli-40-years-since-a-historic-concert/. Mustonen is quoted in Oliver Kautny, *Arvo Pärt zwischen Ost und West. Rezeptionsgeschichte* (Stuttgart: J. B. Metzler, 2002), 118. On the reception of Pärt's work before 1976, see Kevin C. Karnes, *Arvo Pärt's* Tabula Rasa (Oxford: Oxford University Press, 2017), 17–36; and Christopher J. May, "System, Gesture, Rhetoric: Contexts for Rethinking Tintinnabuli in the Music of Arvo Pärt," DPhil thesis, University of Oxford, 2016, 96–170.

3. Mihkelson, "The Cradle of Tintinnabuli." The program for the Tartu concert of October 25 is preserved in ETMM, MO20. In Tallinn, the suite consisted of *Calix, Modus, Trivium, Für Alina, If Bach Had Kept Bees* (*Wenn Bach Bienen gezüchtet hätte*), *Pari intervallo*, and *In spe*. The Tartu program omitted *Trivium* and *Calix*.

4. Toomas Siitan, interview with the author, Tallinn, October 31, 2017 (in English).

5. Arvo Pärt Centre, Laulasmaa (Arvo Pärdi Keskus, hereafter APK), 2-1.10.

6. Alexei Lubimov, "Vremya radostnykh otkrytiy," in *Eti strannyye semidesyatyye, ili Poterya nevinnosti: Esse, interv'yu, vospominaniye*, ed. Georgy Kizeval'ter (Moscow: Novoye literaturnoye obozreniye), 159.

7. For an appreciative assessment of Pärt's tintinnabuli-style music by an Estonian colleague, published shortly after the official premiere, see Merike Vaitmaa, "Hortus Musicus — muusika aed," *Sirp ja vasar*, December 10, 1976, 10.

8. Ilya Kabakov, *60e–70e . . . Zapiski o neofitsial'noy zhizni v Moskve* (Vienna: Gesellschaft zur Förderung slawitischer Studien, 1999), 51.

9. Estonian Composers' Union (Eesti Heliloojate Liit, Tallinn), Pärt file, 1956–78.

10. The program is preserved in ETMM, M238:1/4.

11. *Test pilota Pirxa/Navigaator Pirx*, dir. Marek Piestrak (PRF-ZF and Tallinnfilm, 1979). On Nora Pärt's report to the Composers' Union, see May, "System, Gesture, Rhetoric," 88–89.

12. ETMM, M238:2/62, M238:2/TA.

13. The date and location of the premiere are recorded in Pärt's musical diary of

April 4 through May 20, 1976, preserved at APK, 2-1.10. A manuscript score of *Modus* bearing the heading "Tintinnabulum 1" is preserved in ETMM, M238:2/13.

14. Peter J. Schmelz, email correspondence with the author, January 30, 2017, summarizing Schmelz's January 2017 interview with Lubimov.

15. May, "System, Gesture, Rhetoric," 227n143.

16. Latvian Centre for Contemporary Art, Riga (Latvijas Laikmetīgas mākslas centrs, hereafter LLMC), Lediņš Collection, unlabeled box. As related by an interlocutor, Pärt remembers little of the festival beyond the premiere of *Saara*. Kristina Kõrver, interview with the author, Laulasmaa, November 1, 2017 (in English).

17. Preserved at LLMC, in the folder "Pirmās diskotēkas RPI."

18. Lediņš family/private archive of Lauris Vorslavs, Riga. I am grateful to Lauris for providing me with a digitized copy of the tape and photographs of the reel and its packaging.

19. With permission granted from LLMC, a digital copy of the recording is now archived by the Arvo Pärt Recorded Archive, maintained by Doug Maskew in Tallinn (APRA 0010328: "Mūsdienu mūzikas dekāde" festival).

20. Pärt's discovery of the syllabic method and composition of the *Missa syllabica* are recorded in his musical diary catalogued as APK 2-1.21. On the syllabic method, see Leopold Brauneiss, "Tintinnabuli: An Introduction," in *Arvo Pärt in Conversation*, ed. Enzo Restagno et al. (Champaign, IL: Dalkey Archive, 2012), 122–25.

21. APK 2-1.21.

22. Documented in his musical diaries at APK 2-1.18, 2-1.19.

23. See, for instance, recollections by the violinist Tatiana Grindenko and the painter Ilya Kabakov, in Elena Dvoskina, "Tat'yana Grindenko: novyy put'," *Muzykal'naya akademiya* 2003, no. 2: 51; and Kabakov, *60e–70e*, 51–53.

24. Alexei Yurchak, *Everything Was Forever, Until It Was No More: The Last Soviet Generation* (Princeton, NJ: Princeton University Press, 2006), 112.

25. For a penetrating consideration of the *Credo* scandal, see Toomas Sitaan, "Arvo Pärt—pesni izgnannika," *Muzykal'naya akademiya* 1999, no. 10: 185.

26. The diaries are preserved in APK 2-1.18. Sketches for *Calix* setting the text of the dies irae hymn are preserved in ETMM, M238:2/61; they are described in Vaitmaa, "Arvo Pärdi vokaallooming," *Teater. Muusika. Kino* 1991, no. 2: 22.

27. The program is preserved in ETMM M238:1/4. An archival recording of the concert is preserved at Estonian Public Broadcasting in Tallinn (Eesti Rahvusringhääling), ÜPST-2734/KCDR-1020.

28. Aina Bērziņa, "Jaunie studenti!" *Jaunais inženieris*, October 14, 1974.

29. LLMC, Lediņš Collection, folders "Rokraksti" and "Manuskripti" and notebook "M. Davis."

30. LLMC, Lediņš Collection," folder "Manuskripti."

31. National Archives of Latvia, Riga (Latvijas Valsts arhīvs, hereafter LVA), PA-4263/8/2 (Komsomol protocols of December 16, 1974).

32. LVA, PA-4263/10/1 (protocols of January 26, 1976), PA-4263/10/2 (protocols of September 20, 1976).

33. Asja Visocka, interview with the author, Riga, September 5, 2018 (in Latvian).

34. Preserved at LLMC, Lediņš Collection, folder "Pirmās diskotēkas RPI."

35. Boriss Avramecs, "Neoficiālie laikmetīgās mūzikas festivāli 1976. un 1977. gados Rīgā," in *Robežu pārkāpšana. Mākslu sintēze un paralēles 80. gadi*, ed. Ieva Astahovska (Riga: Laikmetīgās Mākslas Centrs, 2006), 28–29.

36. Donatas Katkus, interview with the author, Vilnius, May 31, 2018 (in English).

37. Preserved at the Paul Sacher Stiftung in Basel, Valentin Silvestrov Collection.

38. Lediņš family/private archive of Lauris Vorslavs, Riga.

39. Quoted in Margarita Katunyan, "Unikal'nyy eksperiment so vremenem," *Muzykal'naya akademiya* 1999, no.3: 3.

40. Vladimir Martynov, "Povorot 1974–1975 godov," in *Eti strannyye semidesyatyye, ili Poterya nevinnosti: Esse, interv'yu, vospominaniye*, ed. Georgy Kizeval'ter (Moscow: Novoye literaturnoye obozreniye), 174–75.

41. Martynov, *Avtoarkheologiya 1978–1998* (Moskow: Klassika-XXI, 2012), 14–15.

42. Martynov, *Avtoarkheologiya*, 22.

43. Martynov, *Avtoarkheologiya*, 23.

44. Martynov, *Avtoarkheologiya*, 31–33.

45. Martynov, *Avtoarkheologiya*, 32. He elaborates the notion of composing with "simulacra" in Vladimir Martynov, *Zona opus posth, ili Rozhdeniye novoy real'nosti* (Moskow: Klassika-XXI, 2011), 17.

46. Avramecs, interview with the author, Riga, November 2, 2017 (in Latvian). Avramecs, "Neoficiālie laikmetīgās mīzikas festivāli," 30.

47. Avramecs describes hearing about reports of an attempted intervention in Anton Rovner, "Riga — moye muzykal'noye prizvaniye," *Muzyka i vremya* 2000, no. 5: 64. The Komsomol archives are preserved in LVA, PA-4263; the Composers' Union archives in LVA, 423.

48. Quoted in Igors Vasiļjevs, "Tiesa," *Liesma* 1988, no. 4: 14.

49. Lediņš, "Vai esmu iegājis vesturē kā mūziķis . . . ," *Padomju jaunatne*, October 21, 1989.

50. Lubimov, "Vremya radostnykh otkrytiy," 156.

51. Reported by Lediņš in Vasiļjevs, "Tiesa," 14; and by Avramecs in Avramecs, "Neoficiālie laikmetīgās mūzikas festivāli," 30.

52. Visocka, interview with the author, Riga, September 5, 2018 (in Latvian).

53. Martin Boiko, interview with the author, Riga, September 3, 2018 (in English).

54. Mustonen recounts the origins of the 1978 Festival of Early and Contemporary Music in Tallinn in Ivalo Randalu, "Muutumised — Andres Mustonen Arvo Pärdist ja iseendast," *Teater. Muusika. Kino* 1995, no. 11: 36–37.

55. Avramecs, interview with the author, Riga, May 7, 2019 (in Latvian); Lubimov, "Vremya radostnykh otkrytiy," 159; Martynov, "Povorot 1974–1975 godov," 174–75.

56. Avramecs, "Neoficiālie laikmetīgās mūzikas festivāli," 30.

.

III

Performance

6

The Pärt Sound

A Conversation with Paul Hillier

The following has been adapted by Paul Hillier and Peter C. Bouteneff from their recorded conversation on September 29, 2016, at Hillier's home in Copenhagen.

Peter Bouteneff. You've conducted ensembles embracing a wide variety of styles and composers and periods. Is there a particular sonic aesthetic that you look for in a Pärt composition, such that you have to instruct your singers in a particular way to approach it? And if so, how might you describe that?

Paul Hillier. I fell in love with Arvo's music the first time I listened to the ECM *Tabula rasa* CD in the early 1980s, and I think all my later work on it is an attempt to relive that experience. The music remains very fresh to me, and it is still very close to my heart. So when I come to rehearse his music with a new group of singers (or a familiar group, for that matter) I want to introduce it to them in such a way that they will understand something, namely, that, although the notes are simple, the music itself has mysteries that must be unfolded carefully. And I don't do this by talking but by handling the music, rehearsing it, performing it. It's important too that they realize I want them to sing supporting the sound with their whole body, even (perhaps especially) when it's soft, and yet also with *fortes* that are not forced. On the one hand it is very precious and delicate, but on the other hand, and often at the same time, it is very strong and powerful. So how to bring those extremes together?

In practical terms there are two things to get to grips with. First, the music must be in tune, and in tintinnabuli music that means the core triad has to ring as true as possible. And secondly the sound must be sustained as a line rather than a series of individual pitches.

Bouteneff. Would you say that within each note, there's a line?

Hillier. Well, you could say there is a line inside each note that leads it to the next. This then turns into a continuity so that the harmonies hold together and flow just like a line of plainchant. Arvo's harmonies are a mixture of consonant and dissonant chords that simply do not function like chromatic harmonies in Wagner, nor do they use the careful preparation and resolution of dissonances in Renaissance music. The nearest similarity I can think of is found in music of the thirteenth and fourteenth centuries.

In the Hilliard Ensemble, when we started singing Pärt, we were coming from several years of singing medieval polyphony — Machaut, Ockeghem, Dufay, and so on — and this background certainly gave us a strong advantage. And the fact that we all shared it meant that we had a collective point of departure when we first started to sing tintinnabuli. The music was new, and it was unusual, but the fact that it seemed so bare did not trouble us: We had sung our Gothic motets, our organum, our conductus, and so Pärt's tintinnabuli made sense to us right away.

Bouteneff. Concerning the "line," for him it's a matter that the measures are each in different time signatures, rather than just having no measures at all, isn't that right?

Hillier. Yes, this is another key element. In the vocal music Arvo used a system of barring whereby one word is a bar. This process has become more variable over time, but in the ur-tintinnabuli period he was very strict about this. It means that there is no regular meter (no dance rhythms therefore!) and no obvious indication of where the main syllabic emphases are. For example, a word like "magnificat" might appear as a bar of 4/4, with the main syllable on the second beat, or "omnipotentem" as a bar of 5/4 with the main emphasis on the fourth beat. So how does one conduct this in a way that is useful to the singers? And remember, it is not just one or two bars that are like this but usually the whole work.

Bouteneff. I was going to ask, do you do a kind of fixed pulse, then?

Hillier. Well, I use the patterns so that we know where we are in the bar, but without placing an ictus on each beat. So for Pärt's music I think of it as a way of subdividing the bar: using the kind of lightly lifted semibeats that allow me to show the full count in any given bar, with a moderate ictus on those strong syllables, while reflecting the whole shape of the word that forms that bar. This allows me to keep the pulse but shape each individual word in turn.

Bouteneff. His assigning one word to a bar — do you suppose that's in part reflecting the priority of the word for him personally?

Hillier. Yes, definitely, and it also represents the way Arvo put the music together. And even though "rationalizing" the barring could make the music easier to conduct from a certain perspective, keeping the words in varying bar lengths retains the original compositional process, and that is surely the stronger perspective. It means that the music hasn't been translated into another more "convenient" notational system but instead actually reflects the way in which the composer conceives the music — or in which he writes down what he conceives — and I see why he would retain that even though it may be awkward to conduct at first glance.

There have been exceptions, for example, the "Credo" in the *Berliner Messe*. This has a lot of text, and, as it happens, the original barring was frustratingly counterintuitive, in the way that the barring worked against the emphases in the text . . . and it gave rhythmic stress patterns that start in one bar and end somewhere in the middle of another bar, maybe two bars later. As this choral work now exists in a version with string orchestra (the original used only organ), the realities of rehearsal time called for a more practical presentation of the accentual patterns, and a revised barring of this movement was subsequently published. This was a compromise that makes sense.

Bouteneff. There are different ways, evidently, of conducting Pärt scores, with the way that they are barred. I noticed this once in a particularly poor performance of *Passio*, that there was a slavish downbeat at the beginning of each measure.

Hillier. Well it really is ingrained in so many of us today that the first beat of a bar should be accented — and the standard conducting patterns only reinforce this idea. This is in contrast to something one learns from singing Renaissance polyphony. In the original partbooks there were no bar lines. There is a meter, but it is merely a way of counting and does not indicate stress. Someone familiar with this older concept is already better equipped to deal with the phrasing of Arvo's music.

Bouteneff. So an ensemble with experience in early music and plainchant and polyphony is going to have a head start, as you say.

Hillier. *Should* have, yes.

Bouteneff. What are some of the particular problems for a conductor of Arvo's music that are distinct from conducting early polyphony and plainchant?

Hillier. We often talk about the influence of early music on Arvo's tintinnabuli music and some of his earlier pieces, but the interesting question might also be: What are the differences?

Bouteneff. Exactly, right.

Hillier. The main difference lies between the nature of Renaissance polyphony and the tintinnabuli style. Renaissance music was essentially
imitative — in other words, you have one voice part which is then imitated (usually at a different pitch) by another voice part. This doesn't
really happen in tintinnabuli music, which is essentially homophonic.
It's based on two notes being sounded together: the melodic voice and the
triadic voice, moving together rhythmically. And whether it uses just two
voices or more, the basic fabric of the tintinnabuli style functions in this
way. Those two voices can be added to (often in further pairs), and Arvo
develops numerous ways of varying and extending the musical texture,
but the essence of tintinnabuli technique lies in this pairing of the two
voices. Very little Renaissance music is like this: As I mentioned, it is in
the medieval period that we find closer resemblances.

In some early medieval polyphony we find examples of note-against-
note composition. A composer takes perhaps a piece of plainchant, writes
the pitches down in a rhythmic pattern, and then adds another part above
it more or less in the same rhythm. The rationale behind the choosing of
pitches for the second part is quite different from tintinnabuli, but the re-
semblance is nonetheless clear.

Another connection to Arvo's music can be found in the development
(around the fifteenth century) of improvising extra voice parts to a given
plainsong, which all church singers had to learn. Once they had their
plainchant by heart, they could improvise further parts (moving mostly in
parallel above or below the chant) to produce a series of what we might
call 6/3 chords or first inversions. This produces a very mellifluous texture
and is often used in composed (that is, notated) music from that period. It
can be found particularly in the English carols of that time, some lovely
ones. Arvo quotes such textures in his Symphony No. 3 (composed in the
supposedly silent years before the invention of tintinnabuli!) and later, for
example in *Da pacem*. Citing Symphony No. 3 is perhaps a rather superficial example of influence but certainly underlines the intensity of his
feeling for early music at a key time in his life.

Bouteneff. Then there are the four-part chorales, and homophony . . .

Hillier. Yes, this is a very different process from the Bach chorale style, actually. I was just thinking about this yesterday in preparation for our conversation, and I looked at one of William Byrd's more homophonic motets,
the *Ave verum corpus*. I asked myself how might Arvo take that melody,
should he want to, and work it in a tintinnabular manner. The answer as
far as I can see is that it doesn't work very well! But still, I took the melody

of the *Ave verum* and put it in a slow triple rhythm and then subjected
that to the descant style that I was describing just now, the fifteenth-
century [style], and you immediately have got a fragment of fifteenth-
century music, using a line of William Byrd.

I'm not sure how revealing this is, but, having looked at the fifteenth-
century approach and also [having] played with the tintinnabuli approach using Byrd's melody (the soprano or cantus part), I then went
back to the Byrd original, and of course what immediately stands out is
the bass part, which has a completely different job to do, in Bach's chorales, but also in the Byrd motet. The Byrd is really made by what the
bass line does against the melody, as is the case even more so in a Bach
chorale. Of course the other parts are important, but with a chorale one
would usually take the given melody and begin by composing a bass part
to it, so that there's a real dialogue between the two outer voices.

Bouteneff. One might factor in rhythm as well. Isn't the bass part often a
kind of a pulse?

Hillier. Yes, but that role of the bass part doesn't really exist in tintinnabuli
music. It may be a melodic line, it may be a tintinnabuli voice, but it's
not an independent bass part. The only thing Arvo does that comes close
to that is to use a pedal note from time to time, but that's an ancient technique coming from somewhere quite different. It just illustrates how foreign the tintinnabuli style is to such well-known kinds of "early music" —
sixteenth-century polyphony and Bach — in terms of how the notes are
put together. It's a different aesthetic altogether.

Bouteneff. I want to come back to a word that you used very significantly,
since we're especially focused here on the subject of sound, and that is
the word "ring." When you spoke of the importance of intervals being
properly in tune, you said that this is heightened when it comes to that
triad voice — and of course performatively that's challenging because
when you combine that with the melody voice you're going to end up
with a lot of both consonant and dissonant intervals — but if you'd describe that "ring" a little and what exigencies that puts upon a choir and
its director. You've mentioned intonation, of course . . .

Hillier. Intonation is one of the most important things in this music, as is
sustaining the sound so that the full nature of each chord is fully registered, whether or not it contains dissonances. And because the same
fundamental triad is being sounded much of the time, then you do feel
it ringing throughout the music, even though at some moments there
are silences. When the music picks up again it's as if the sound has been
there all the time but simply passes out of our hearing for a moment. It's

like when you re-touch a bell that's already been sounded: It continues to ring, but then you might want to just give it another ping, just to bring the sound out again, and that in a way is what happens in Arvo's music.

Bouteneff. How do you give that "additional ping" with a choir? What do you tell them to do?

Hillier. Well that part of it is composed into the music, and the triad may move around, up and down, but it's the same basic triad, always being tested and rubbed up against dissonant elements in the M-voice. But the core sound is constantly being refreshed as the music continues.

Bouteneff. That's what gives it that additional "ping."

Hillier. Yes, and naturally it depends a lot on the acoustic space in which you're performing. In a sense the building is also performing the music, and this is something the singers need to understand.

Bouteneff. Are you, then, listening for that, again, that "ring" in whatever space you're performing and recording in?

Hillier. I am, but of course you don't always get to perform in ideal acoustics. The sound is so different from place to place, and that makes it all the more important to have singers who understand what they're doing, and can sustain the sound — and not be inhibited by a dry acoustic. And at the same time they must sing within themselves rather than strive to make more noise in a mistaken attempt to fill the room. It's still possible to perform, and to perform well, in a dry acoustic, but I would be the first to admit that it's hard work, and the overall experience simply isn't quite as strong. The nature of the venue really does matter.

Bouteneff. What about the role of vibrato, or nonvibrato?

Hillier. Ah. Well, there's nothing wrong with a small natural vibrato, and sometimes I ask a singer to use that if their sound is a little tight or colorless. But the kind of vibrato I dislike is the kind that's trained into a voice by a singing teacher, to the extent that it becomes pitch wobble. First of all, the pitch wobble inevitably clouds the whole issue of what chord we're listening to. This is something that can hurt all kinds of music, especially polyphony and contemporary music of the harmonically more static variety. So there is the desire simply to keep the sound clear enough so that the nature of the chord is honored, is completely there.

Bouteneff. So vibrato can pertain not only to pitch but also to dynamics.

Hillier. Yes, and the other thing with vibrato which can be problematic is that it pulls the ear towards it. It's a subjective way of singing that is too . . . well, if I sing "Ah" [without vibrato] and then if I sing "Ah" [with vibrato]. As soon as I put that kind of variation of pitch into the middle of the sound, then the ear can't help but latch onto it.

Bouteneff. Yes. I heard almost a full minor third within that last vibrato!

Hillier. Now, a lot of singers who have been trained in the modern classical way *do* use vibrato. Or perhaps they think they're not using vibrato
but can't stop themselves from introducing vibrato towards the end of a
note . . . every note longer than a quarter note. And that quickly becomes
a mannerism which I find irritating, because for me it destroys the real
nature of the music and creates an all-purpose *espressivo* that in fact ne-
gates any kind of true expression.

So that's the problem I have with vibrato. It blurs harmony, it's overly
subjective, and it prevents me from hearing the music. Vibrato should
only be used as one among many elements of embellishment — to be
used sparingly and with a sense of musical decorum!

Bouteneff. So it seems to me there's a real primacy of pitch, and that anything that might interfere with the pitch, such as a pitched vibrato, or any
other factor, would be problematic?

Hillier. There is a small area where there's a very light natural fluctuation of
pitch, but that's all.

Bouteneff. One more thing, perhaps, related to vibrato: I was thinking of the
Passio, and you have, of course, the soloists, you have the evangelist quartet, and then you have the Turba choir. It seems to me that each has sort
of different mandates, musically speaking, or in any case, I'm hearing a different vocal aesthetic, as it were, even between the soloists — Pilate, who is
forced to sing more intervals, of course — but the desirability of something
like vibrato or dynamic manipulation seems to me different in whether it's
a soloist — and *which* soloist — evangelist, and Turba. Is that accurate?

Hillier. Up to a point, yes. The two purely solo roles in *Passio* (Christus and
Pilatus) *can* be sung by singers using a moderate vibrato without spoiling
the effect — but they certainly don't *have* to be. The Evangelist quartet
[soprano, alto, tenor, bass] are mostly singing as a vocal ensemble, and we
need to hear clarity in their pitch, and they each need to sing with very
much the same kind of voice production. I would say that the same is
true, by extension, of the Chorus who sing the other roles, including the
Turba or crowd. The instrumental quartet that accompanies the Evangelist quartet need to integrate their sound with that of the singers, and this
naturally includes the restrained use of vibrato. That leaves the organist
with a range of interesting responsibilities, and sometimes Chris Bowers
Broadbent (the organist I've used for *Passio* almost every single time I've
done it) uses an organ vibrato stop — the vox humana! — to underline the
ambiguities of Pilatus's position. So there are all sorts of coloristic free-
doms to be found even in a world without vibrato.

Bouteneff. Back to a classical sort of tintinnabuli two-voice situation, and again perhaps with that "ring" in mind . . . do you ever have to instruct the M-voice or the T-voice to be more dominant, louder, for example, quieter, or do they have to be exactly as if they were two simultaneously struck keys?

Hillier. Arvo often used to say, pointing to the M-voice, "this is the line we must hear."

Bouteneff. So that's what he prioritizes, the M-voice?

Hillier. The melodic voice is the one that is irregular, changeable, even, so to speak, subjective, yet it is embraced and held in place by the T-voices, which are the triads. Its line should be audible and should not get lost inside the sound of the triads. For example, in the "Exordium" and "Conclusio" of *Passio*, the M-voices are scales, in the first descending, in the second moving in both directions. Arvo wanted this scale to be perceptible.

Bouteneff. As a melody, as a line.

Hillier. But not as a solo-plus-accompaniment. It's not that. The total sound should still be in balance. I think this is something too that the performers need to understand, so they can adjust their volume and articulation, probably only by very small degrees, to achieve what's needed.

Bouteneff. [looking through some assembled scores including the original Estonian editions of *An den Wassern zu Babel saßen wir und weinten* and *Sarah Was Ninety Years Old*, which at that time were titled *In spe* and *Modus* — See Figure 1 below.] Can we talk about notation for a moment?

Hillier. I just love this spare way of notating things.

Bouteneff. It's pretty unique, isn't it?

Hillier. And that survived even when he went with Universal Edition and started using more regular notation. The score was still pretty bare and without expression marks.

I've noticed that he's using those marks more and more, and I think it's a shame, a little bit anyway. Anyone who understands the music is going to get there anyway, and anyone who doesn't is probably going to exaggerate the markings because that's what happens: One singer's *f* will be another's *mp*, stress marks will receive heavy accents, breaths will become short pauses. And the presence of expression marks only reinforces the notion that their absence (for you cannot mark everything, although some composers have tried) implies a totally passive and deliberately level style of playing.

Bouteneff. Yes, I'm glad you raised that . . . So you're saying that, with time, he increases the amount of instruction he gives within his scores?

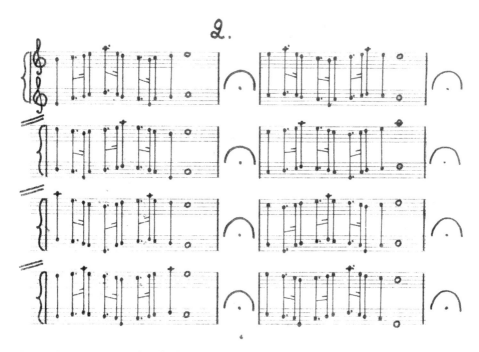

Figure 1. Manuscript page of *Modus* (*Sarah Was Ninety Years Old*), 1977. Courtesy of the Arvo Pärt Centre, APC 2-2.4.159 p4.

Hillier. Well, it seems to have been the case that over the past decade or so this has steadily increased. Perhaps it's because, knowing that his music is being performed by more and more people, some of whom may choose to do just one piece here or there without getting to know and under-stand his oeuvre as a whole, he therefore regards this as a practical neces-sity. I don't know.

Bouteneff. Is it unnecessarily constrictive?

Hillier. Well, to my mind it is. We can compare it with, let's say, fifteenth-century music, when expression marks did not exist. There was no need for them. That doesn't mean there's no expression, but it comes from knowing the style, from living with the music, and using your common sense as well.

Bouteneff. That's interesting, because I've heard Arvo say that the more and more his music gets performed (unlike in the early '80s), the more there's a kind of a "gestalt" that gets passed on, ensemble to ensemble: "Okay, this is a Pärt piece. We know basically what we're striving for," and in a way that would seem to argue for less instruction, and yet, on the other hand, there's more.

Hillier. Maybe I'm wrong in my observations, but I think that's what I see.

Bouteneff. I think there could be both things going on at once.

Hillier. There is one perspective that may be relevant. In recent years Arvo has not stopped using tintinnabuli altogether, but he has used it with increasing flexibility, so that sometimes he even steps outside of it, keeping the tintinnabuli technique as one element amongst others. And perhaps the natural implication of moving in this direction is to feel the need for more interpretative directions. Obviously my problem is that one of the things that first strongly attracted me to Arvo's scores was the absence of expression marks!

Bouteneff. You don't have to answer this if you don't want to, but are there composers that you know who are more and less willing to cede that kind of interpretive freedom to ensembles? I can certainly understand the impulse to guard control of the piece, but . . .

Hillier. Yes there are, and on balance I've always been drawn to music that leaves the performer alone to get on with it and play it. I think the performer's role in music is not to serve the composer; it's to take possession of the music and make it your own.

Bouteneff. Especially before major performances of his work, Arvo often appears at rehearsals, and there are several video-documented moments of his sometimes extremely physical involvement in the rehearsal. He'll speak in metaphors, sometimes accompanied with gestures, or maybe he will tiptoe across the stage or do these almost comical things. What's that like? What is he trying to do?

Hillier. He used to do it with us as well. I can remember, for example, when he took a piece of paper and tore it up into little bits and just let it float down through the air. He was showing us how the notes should sound if we used our imagination in that way. It's an attempt to suggest some aspect of the sound in a physical motion, and it can be very helpful.

Bouteneff. Does it ever run contrary to what's written as a notation on the score?

Hillier. Well, no, because in a sense a score can't do things like that anyway, unless it's a graphic score. (And by the way, there are graphic elements scattered throughout the earlier Pärt scores, especially those designed and printed in Estonia.) But, of course, when a "serious" composer performs these little actions during a rehearsal, he's revealing a side of his personality that is not obvious from the music, and it's endearing; the singers and players relax and enjoy the process more — and perform better!

Bouteneff. I guess the score and those additional instructions he writes on the score in the notations are one way of, as it were, ensuring continuity, ensuring that he gets what is desired. It seems, in a kind of a de facto way, that some recorded performances become authoritative, and many of those, especially the ECM recordings, are maybe even intended to carry that status.

Hillier. I think Arvo was always very conscious that this was the case and felt that it should be so. Saying not "You must do it like this" but more simply "This is the way I like this piece to be done."

Bouteneff. So that that would become the "reference work."

Hillier. As you can possibly surmise, I feel there's more than one answer to these questions. But I've noticed that Arvo uses not only *forte* and *piano* and so on but now also crescendo and decrescendo markings and commas and breaths and all sorts of things, and usually I reckon that the music doesn't need them.

Bouteneff. Does he ever say something more interpretively, you know: "Like a child laughing" or something like that?

Hillier. No, I don't think so, but compared to the early scores, things are now quite different. No, the gestures and comedy are wonderful, and they really do help, at several levels. People also realize that he's human, that he has a great sense of humor and of timing . . . and not just in music. It's the use of metronome speeds, dynamics, and all that paraphernalia which I regret.

Bouteneff. We've been talking thus far largely about what the exigencies of his music require of a director, of an ensemble and performers, so it's the influence from that direction, which is the typical direction, that the composer and his style and his music will affect what an ensemble does. There's also the reverse, though — the ensemble affecting the composer. And I remember being very moved when I first read these words in an interview with Arvo and Nora together, where they describe those rehearsals that you mentioned at the very beginning of this conversation — and I quote: "I remember that we were speechless when we heard those rehearsals. They sang with absolute perfection and showed enormous sensitivity to the style of Arvo's music in his choice of registers. We were moved to tears, overjoyed to have found people that were suited to this music in every way." So at first there's just this level of appreciation . . . just "that was great," but Arvo then adds: "We had never heard such pure fifths and thirds," so the intonation issue within tintinnabuli, and then, significantly, he says only after he heard the Hilliard Ensemble perform *De profundis, Cantate Domino, An den Wassern* . . . : "Only then did I become aware that I had made the right choice with my compositional technique and I understood that this type of music was really viable."[1]

There, the performance is informing — or in this case confirming — for the composer a whole *via*, a whole way. So not only is that a remarkable statement of what your ensemble was able to do, but it's a case where the ensemble had an influence on how the composer works.

Hillier. I don't know if it affected the way he works, but perhaps it encouraged him to go on working in the new direction he had taken. It was confirmation. All composers need to hear their music done in a way they can recognize as right, or at least on the right track. In the '70s, let's say, and into the early '80s, Arvo really was out on a limb, doing something that nobody else was doing. There were parallel activities elsewhere, but I don't think he was aware of them at the time except as distant murmurs. I'm thinking of Cage and the so-called New York School, Fluxus, the earlier years of minimalism, and the European "new simplicity." So, given the state of new music just then, which as we all know was essentially one of complexity, there's no question that he was somewhere else! And in that somewhere else — Soviet-era Estonia — he had been strongly criticized, first of all for writing modernist music and then, once that had become accepted, strongly criticized for turning away and writing this kind of music. People thought he was mad.

Bouteneff. That's right.

Hillier. So it's not surprising that, when he heard it being sung more or less the way he needed it to be sung, it was going to make a big impression.

Bouteneff. But it seems to be almost beyond that. It's not like: "Ah, that's what I was looking for." It's almost approaching: "I didn't even know how powerful this could sound . . ."

Hillier. But in a way that's exactly what I mean. There is no substitute for the actual sounding music, because then something else is going on. We can sit and look at a score and have a sense of what's there and get excited about it, but when you actually make the sounds, then everything changes, and it doesn't really matter what the music is, and especially when it's something that is so new and seemingly empty, of course he needed to hear it come together.

But as I said before, we could sing it, because by luck we came out of the traditions that he needed, which is to say, singing early music in the way that we were singing it: in tune without vibrato, with agile rhythmic articulation, and in long sustained lines. We certainly weren't the only people doing that, but it was in England in the 1960s and 1970s that this kind of singing really took root. There were other groups besides the Hilliards, but not so many, and the various groups that existed mostly used the same collection of singers in varying combinations! I remember very clearly that we were learning from one another: We didn't know where we were going exactly, but we recognized each turn as we came to it. We then brought that experience to bear on singing Arvo's music, and that's how and why it came together in the way it did.

Bouteneff. Wonderful. As you said, it comes down to the sound, doesn't it? Now, a further thing about your contribution: Apart from being so instrumental in recording his music early on and performing it in this authoritative way — in the sense that these became the authoritative recordings — you wrote the book that launched the terminology of M- and T-voices, etc. Would you say something about the relationship between a written analysis of how tintinnabuli music works, the tensions and resolutions, the way that air flows with a minor and with a major triad, for example, overtone series, etc. — the relationship between that and the "ring" that you speak of and the more visceral, as it were, impact of it? I suppose one could ask that question about any analysis of music versus the experience of it, analysis of art and the experience of it, but in the case of tintinnabuli, can one relate these two to each other; can one bring them into some kind of fruitful dialogue?

Hillier. Difficult. You're basically asking me to explain the nature of music and why some music —

Bouteneff. Go for it!

Hillier. Well, I don't think I can. Nobody really can explain what makes great music great. They can show what's going on in it, but our interest is only aroused in the first place because of the sound itself. We don't analyze a piece of music and then listen to it for the first time. Usually we hear something or we listen to something or we play something, and if we think it's fantastic, *then* we want to find out why and how it's put together. In the process of examining how something is made, you learn a lot about the nature of the music, and you put yourself in a potentially privileged position to perform that music better, because you understand it not only from the intuitive side but also from the other more analytical angle at the same time. It's not that one has to be able to write mathematical analyses of which we see so many examples.

Bouteneff. That has been done, yes.

Hillier. And they can help. But then I think they have to be set aside so that you can live in the music from one moment to the next, strengthened by the knowledge you've gained about the work as a whole. You go for a drive and enjoy the scenery, but you know all the time where you are and how you're going to get to your destination.

When I studied the various interlocking processes in *Passio* for my

book and traced out a diagram of how the Evangelist music was con-
structed (the four voices and the four instruments together), I remember watching with great pleasure as the pattern emerged in front of me.
It didn't tell me a thing about how to sing or play the notes, but it meant
that I always knew where I was at any given moment in the piece, and
that kind of structural awareness is very important.

Bouteneff. So it's really a conversation between experience, analysis, and
again experience.

Hillier. Exactly, and the sound comes first.

Bouteneff. Is it possible to say just a little bit more about that experience.
There's the sound experience, then the analysis, and you spoke about
how the analysis can then influence the performance or even the listening.

Hillier. Some people, as listeners, may say they just want to let themselves
go and soak up the music from one moment to the next. I totally agree
with them! That's how I listen too. But for a performer that moment-to-
moment process has to be supported by an understanding of the whole
work, as fully as possible — in fact with great works of music you never
reach the end of learning things about them. That kind of knowledge,
however, is available to everyone, and you don't need a music degree to
obtain it. But, again, the performer is in a different position — and has a
different kind of responsibility — but whoever you are, then a deeper understanding of the form of a musical work is going to deepen the pleasure
you get from it.

Bouteneff. As you say, perform intelligently, or with intelligence.

Hillier. Yes, and with vocal music there's the added dimension of the text.
My approach is to examine the text by itself away from the music — not
just what do the words mean, but as with the music, how is the text actually structured? Then I can go back to music + text and see how the
composer has used the text form, in some cases reflecting it directly in
the music, in other cases ignoring it altogether, breaking it up and con-
structing a new and purely musical form. The truth of course is usually
somewhere in the middle. I also like to reverse the process and look at the
music without the text at all; that also can be very revealing. I find these
processes extremely informative (in fact, it saves time!) and also fascinating. It's a kind of rough and ready rhetorical study, if you like. The listener doesn't need to bother with all this, because when performers prepare themselves in this way, then the process informs the performance,
which in turn gives it over to the listener.

Bouteneff. Right, and then the listener reaps the benefits in a more subliminal way.

Moving on, there's been so much said about *silence* in Arvo's music. And apart from ways in which we might say that his music is "characterized" by silence, there are also sometimes pauses, rests, longer or shorter, that are literally unfilled with sound. So there is, ironically, a sonic impact of silence on the listener, then?

Hillier. We have to be careful here, because as we said earlier there are so many different acoustics. As with plainsong and certain kinds of polyphony, so with Pärt: The acoustic becomes part of the sound, not an add-on.

Bouteneff. I think of the beginning of *Miserere*, which has so much space between the notes. And you're right: The echo/reverberation of a good venue will have it such that there's no actual dead air, as it were.

Hillier. Yes. And at the very least, in different acoustics those silences will differ in length.

Bouteneff. Because you're waiting until the . . .

Hillier. I don't think I'm waiting for anything. But I don't count the silences precisely anyway, not in a work I've performed before. In a concert I'm waiting for it to be intuitively right to move on — but with intuition based on earlier experiences of that same work. However, this doesn't stop me from taking risks with silences even in a dry acoustic to see what can be achieved, how long can they endure and remain "musical." It depends on many little things, including the amount of light in the hall, the temperature(!), and the audience . . . are they restive (and thus spoiling the silences for one another) or are they with us. And if they are not with us, I will still be listening and trying to create an ambience in which the silences remain meaningful. Hurrying never works!

Bouteneff. And there, too, just as there has been an increasing awareness on the part of performers of Pärt's music — awareness of what to expect from it — listeners, too, now come expecting something from a Pärt work, knowing that it's not going to be a full-on entertainment, as it were, but there's going to be some "breathing."

Hillier. And with regard to those breathing silences, what matters is how you play at the end of the silence, how you start the next phrase. There are many different ways of playing the next phrase after a silence, and the silence (the acoustic again) will affect how you play the next note. Now, it may be a loud note that needs to go *ping* suddenly, but more often it'll be something else . . . Actually John Cage came up with a very good way of describing that kind of thing. About one of his pieces where there are many silences he said: It should be as if the note is "brushed into existence"—which I think says it all. Again, this is a little bit like Arvo's gestures. You understand exactly what Cage means, yet there is no brush, only a feeling.

Bouteneff. One thinks of the Japanese or Chinese calligraphy.

Hillier. Exactly, especially in Cage's case.

Bouteneff. In the case of Chinese and Japanese traditions, it's said that the calligraphy begins when the brush is in the air, before it even descends, and then as it descends, it will begin making its visible mark and then it continues even as the brush is lifted off the paper . . .

Hillier. Yes, that's a very good way of suggesting how to approach Arvo's phrases.

Bouteneff. There was a term I think you used at the end of your book to describe his music, and I think it's "abstract tonality," which I don't know if you mean it as something of an oxymoron or in an ironic sense or . . . of course the word "tonal" certainly has come to mean a lot of different things . . .

Hillier. Yes, I'm not sure if it's a good phrase or not, but I do mean it without any form of irony. In Western art the development of perspective called forth painting that was essentially representational, a picture of the visible world. It's only during the past hundred years that we've experienced first the expressive distortion of reality leading eventually into fully abstract (nonrepresentational) art and other versions or subversions of perspective. Similarly in music, though somewhat later in time, composers developed the possibilities of functional tonality by building on the tension and release of dissonance into consonance, and then extended its reach through the use of extended chromaticism, leading us (modulating) into new tonal areas, and eventually to the use of all twelve pitches to avoid tonality altogether. In minimalism we have a kind of music that is obviously tonal but which does not use structured chromaticism to move, over long periods of musical time, from one tonal area to another. That kind of emotional push-pull was abandoned in favor of a more static harmonic field. It doesn't mean there were no dissonances (listen to Andriessen, for example), but these become part of the harmonic field. And so if (perhaps a big "if") we can view functional tonality as something akin to representational art, then minimalist music (so called) is abstract tonality.

Bouteneff. Just as you're talking, the piece *Solfeggio* comes to mind, which is, innocently enough, an ascending diatonic scale, yet it reveals . . .

suddenly you're hearing minor ninths and hearing these crazy intervals,
but they're right there, hiding in plain sight within the diatonic scale.

Hillier. I think it's a brilliant piece, and it's very interesting that it was written so long back, before he even began to explore tonal music as such.

Bouteneff. But those simple scales are there in *Passio*, as you said, and in so
very many of his compositions. They are everywhere, deceptively simple.

Hillier. But of course what he does in *Solfeggio* is to elongate some of the
pitches and just leave them there, so that various dissonant clusters are
built up with, so to speak, a purposeful lack of direction. It's a tonal world
where you have only M-voices, no T-voice to get them out of trouble!

Bouteneff. We've been kind of going in and out of the directly sonic implications of Arvo's music . . .

Hillier. I find that, just as Arvo's whole oeuvre is unmistakably itself, so I
find that the best pieces of his, the ones that really hold our attention
time and again, also have their own private sound-world. I'm thinking, for
example, of *Fratres*. There's the "*Fratres* sound" that no other piece, even
in Arvo's work, imitates, and I think it's one of his greatest pieces. There
are other pieces of which I would say the same thing. They are unique.
Even a piece such as *Spiegel im Spiegel*, which has become almost over-
used in film soundtracks — even so, I find it very special.

Bouteneff. For a while it seemed like it was on *every* soundtrack!

Hillier. And like several of his instrumental works, it's basically a stepwise
melody and some slowly arpeggiated triads, easily imitated perhaps. Yet
it has that indefinable quality that keeps it internally fresh, and the piece
has its own sound. Even in all those films!

The same can be said of certain of the choral works. I really love them
all, but there are a handful which in my opinion stand out, and the reason for that is their complete individuality within the oeuvre. I won't list
them, just mention one, *Magnificat*. A masterpiece.

It is interesting, however, that in more recent years there have been
works that move away from the general picture and leave you wondering
for a moment . . .

Bouteneff. And you say, "Is that Arvo?"

Hillier. Yes, though the question usually answers itself quite quickly. But
with *Fratres* you enter a specific world and stay there for the whole piece.

Bouteneff. If we could get into just one more area of exploration . . . and that
is the eponymous *bell* in tintinnabuli. If I even say as much as that — the
bell — how would you begin to take that up . . . And you spoke of the "ring"
that one looks for in a tintinnabuli composition and its performance . . .

Hillier. It's a topic that fascinates me. When studying Arvo's music in detail for my book, I realized I had to find out more about the Russian bell tradition. Where did Arvo's ideas come from? I had seen photographs of Arvo from the early years with bells in them — there's one where he's holding a bell up and looking at it meaningfully — so I knew there was something there that might help me understand more about the tintinnabuli muse. The Russian *zvon* tradition — as you will certainly know — is a world of its own, utterly distinct from Western European bell ringing. But the recordings of Russian and Greek Orthodox bell music that I managed to acquire introduced me to something very new, and this widened my frame of references for thinking about Arvo's tintinnabuli music.

But then in the late '80s Heinz Holliger wrote a piece for the Hilliards in which a Japanese *rin* bell was used — this is a fairly small bell, usually made of brass I think, that has a very pure sound and which, if left alone, rings for a surprisingly long time, only disappearing very gradually. And just when you think it has finally fallen silent, if you go over to it and put your ear close, you hear that it's still ringing. Who can say when it really stops! (This lovely phenomenon reminded me, by the way, of the concept underlying Gavin Bryars's *The Sinking of the Titanic*: the idea that the music played by the band as the ship sank into the sea was still vibrating somewhere at the bottom of the ocean a hundred years later.) A year or two later, on a tour of Japan, I found a Buddhist shop that sold such bells and bought a couple to take home, so now I had many opportunities to explore the nature of this sound, and inevitably it fueled my thoughts about the connection between bells and the sound of Arvo's music. I concluded that there was something crucial to understanding the tintinnabuli aesthetic in the way the sound of the bell disappears but doesn't disappear: It's still there.

Bouteneff. There, but not there?

Hillier. Or, not there, yet still present.

Note

1. Enzo Restagno et al., *Arvo Pärt in Conversation*, trans. Robert Crow (Champaign: Dalkey Archive, 2012), 49–51.

7

The Rest Is Silence

Andrew Shenton

> When the inexpressible had to be expressed, Shakespeare laid down
> his pen and called for music. And if the music should also fail?
> Well, there was always silence to fall back on. For always, always and
> everywhere, the rest is silence.
>
> —ALDOUS HUXLEY

Writing in 1931, Huxley points out that silence is the closest we have to being
able to express the inexpressible, better even than music.[1] He notes that we are
surrounded by silence, perhaps not in a literal sense, but in a profound spiritual sense, in which silence can mean both the absence and presence of the
divine. Both are important to Arvo Pärt, who is renowned for his use of silence
in the works composed after 1976. Apart from noting its frequency in his work
and its structural and spiritual efficacy, the subject has not been explored in
practical detail. This essay analyzes the spectrum of silences in Pärt's music.

For our purposes, silence is being considered in two broad categories, first,
as a metaphorical, theological, or spiritual phenomenon; second, as a practical, physical, and embodied phenomenon. The taxonomy and examples provided here demonstrate their interconnectedness in Pärt's music.

Early in his career Pärt noted the importance of silence in an often-repeated
quotation: "I have discovered that it is enough when a single note is beautifully played. This one note, or a silent beat, or a moment of silence comforts
me."[2] This statement has generally been taken as a whole, but it is interesting
to focus on the two clauses that concern silence. Pärt notes that a silent beat
or a moment of silence comforts him, thereby suggesting that silences, be they
long or short, are consolatory. These personal reactions to silence have reli-

gious overtones for Pärt, but the choice of wording is important. He is simple, direct, and unequivocal in stating a deep, profound truth about his desired soundworld. Pärt also states that a single note comforts him, suggesting something meditative and beautiful in this single pitch that is unobtainable in complex rhythms and harmonies. After all, what is a single note but the minimal interruption of silence?

The trope of the importance of silence in Pärt's music starts in the 1980s and stems principally from early interviews and reviews of his work. These include discussions by Paul Hillier that focus on the hesychast tradition in Orthodox Christianity and on descriptions of silences in his music up to 1997 and a short essay on performance by the conductor Andreas Per Kähler, which suggests that the aesthetic of Pärt's music after 1976 requires a new approach by performers.[3] Some authors, such as Leo Normet, use silence as a general theme for an aesthetic and theological point of entry into analysis of the music.[4] Others have, with more or less specificity, engaged in the use of silence in the abstract (Helbig) or in the context of analysis of individual works (Cobussen, Vuorinen).[5]

The most comprehensive theological analysis of Pärt's use of silence is Peter Bouteneff's *Arvo Pärt: Out of Silence.*[6] Bouteneff covers several aspects of silence as it pertains to Pärt's life and music: silence born of crisis and suffering, as absence and presence, as multivalent and productive, and as part of the Orthodox Christian tradition.[7] He discusses the inherent contradictions in the silence of God's absence and presence and the notable silences recounted during Jesus's life. Bouteneff also expands on the notion of purification by stillness (*hesychia*) and the connection of stillness and silence. Taking examples from a broad range of theologians and Christian writers, Bouteneff discusses different types of silence: silence that emanates from sound and turmoil, or that comes from the awesome presence of God, or that is cultivated as an inner state or temperament, or that produces sound.[8]

Bouteneff has made a case for an (Orthodox) Christian approach to understanding Pärt's silence, but in order to account for the extraordinary popularity of Pärt's music, we can invoke other relevant practices of the Orthodox Church and broaden our understanding to a level that surpasses institutional religion.

Silence as Theology

Pärt acknowledges two kinds of silence in his personal theology:

> When we speak about silence, we must keep in mind that it has two
> different wings, so to speak. Silence can be both that which is outside

of us and that which is inside a person. The silence of our soul, which isn't even affected by external distractions, is actually more crucial but more difficult to achieve.[9]

Pärt recognizes silence as an antipode to sound (silence outside of us) but, more importantly, as a theological concept (the silence of our souls). It is the second of these that interests him most, and his success at finding space for this kind of silence in his music resonates with many people.

Since Pärt is an Orthodox Christian (he entered the Russian Orthodox Church in 1972), it is useful to invoke doctrines and praxes of the Christian Church in assessing his theological use of silence. As Bouteneff notes, "The God of Judeo-Christian tradition is inconceivable and indescribable, and must be approached in silence and darkness — the early Christian writers called this *apophasis*. Yet it has been given to humans to say something, to him and about him, which they called *kataphasis*."[10] For Pärt, *kataphasis* is evident not only through his texts (which are almost exclusively religious) but also through tintinnabulation itself, which manifests some traditional approaches to a deeper communion with the divine.

The hesychast tradition in Eastern Orthodoxy, which upholds silence as the culmination of one's salvation and ascent into the eternal fellowship of the Trinity, has been discussed by scholars in relation to Pärt's music, and its value to Pärt is clear.[11] Hesychasm is a prayer tradition that invites people to retire inward and keep stillness in order to experience the divine. As Hillier notes, "As early as the fourth century it was used to designate the state of inner peace and freedom from bodily or mental passion from which point only one might proceed to actual contemplation."[12] This description is fitting for Pärt in two ways. First, because Pärt describes his own personal experience of silence and stillness as a part of his creative process; and second, because his music can function as the action of hesychasm on listeners, bringing them to a position where they may be able to reflect more deeply on the text they have heard or contemplate a more general experience of the divine.

Pärt admits the utility and power of silence as part of his creative process: "On the one hand, silence is like fertile soil, which awaits our creative act, our seed. But on the other hand, silence must be approached with a feeling of awe."[13] He also confesses how his music is born out of deeply religious silence:

> My music has emerged only after I have been silent for quite some time, literally silent. For me, "silent" means the "nothing" from which God created the world. Ideally, a silent pause is something sacred. . . . If someone approaches silence with love, then this might give birth to music.[14]

These quotations indicate that Pärt is aware of multiple values for the silence he uses. The quotation also points to two other theologies that are relevant to the discussion, particularly as it relates to his use of "comfortable" silence and "the silence of our souls." First, *kenosis*, the doctrine that Christ relinquished his divine attributes so as to experience human suffering. In the Christian tradition this doctrine is interpreted as self-emptying to become receptive to God. In the Orthodox tradition kenosis is possible only through humility and entails continual *epiklesis* (invocation of the Holy Spirit) and self-denial of one's own human will and desire. The sacred silence Pärt keeps before composing (the "nothing from which God created the world") demonstrates this self-emptying in order to be receptive to the Word.

The goal of kenosis is union with God, and this goal can also be achieved through *theosis*, the process of becoming holy by grace. This in turn is brought about by catharsis (purification of mind and body). Ascetic practices, then, usher in a kenosis similar to that of Christ, by which people empty themselves to be filled with the Holy Spirit. As I have noted elsewhere, the drawing of music out of silence, coaxing the audible from the inaudible, is a perfect illustration of kenosis in action.[15]

Silence as Spirituality

These religious contexts are useful to listeners who are Christian and may provide useful background for those who are sympathetic to Christianity, but how do we interpret the use of silence in a nonreligious context? One approach is to consider contemporary notions of spirituality. In her 1966 essay "The Aesthetics of Silence," Susan Sontag suggests that "every era has to reinvent the project of 'spirituality' for itself."[16] Her definition of spirituality is broad and inclusive: "plans, terminologies, ideas of deportment aimed at resolving the painful structural contradictions inherent in the human situation, at the completion of human consciousness, at transcendence."[17]

Sontag describes the "leading myth" of art as the "absoluteness of the artist's activity," by which she means free from imperfections and complete.[18] She goes on to define a "newer myth," which "installs within the activity of art many of the paradoxes involved in attaining an absolute state of being described by the great religious mystics," noting that the

> activity of the mystic must end in a *via negativa*, a theology of God's absence, a craving for the cloud of unknowing beyond knowledge and for the silence beyond speech, so art must tend toward anti-art, the elimination of the "subject" (the "object," the "image"), the substitution of chance for intention, and the pursuit of silence.[19]

Sontag's prescient remarks provide a way of understanding both Pärt's music and its popular appeal.

Sontag notes, "One must acknowledge a surrounding environment of sound or language in order to recognize silence."[20] Pärt's mastery is that through tin-tinnabulation (and to some extent in the pre-tintinnabuli works) he found a way to do this in a personal Christian context without this context being an impediment to appreciation of the artwork by non-Christians.

Again, it is Sontag who illuminates an effect Pärt's music has on audiences: "The notions of silence, emptiness, and reduction sketch out new prescriptions for looking, hearing, etc.—which either promote a more immediate, sensuous experience of art or confront the artwork in a more conspicuous, conceptual way."[21] She also notes, "Perhaps the quality of attention one brings to bear on something will be better (less contaminated, less distracted), the less one is offered."[22] This emphasis on the listening experience is an interesting point of comparison with Pärt, who requires active listening for even the simplest of his pieces. Sontag expands on her point, suggesting:

> In the light of the current myth, in which art aims to become a "total experience," soliciting total attention, the strategies of impoverishment and reduction indicate the most exalted ambition art could adopt. Underneath what looks like a strenuous modesty, if not actual debility, is to be discerned an energetic secular blasphemy: the wish to attain the unfettered, unselective, total consciousness of "God."[23]

In Pärt's aesthetic, in which he has intentionally adopted the "way of reduction," he has managed to capture the attention of his audiences.[24] What, then, are his methodological uses of silence?

Categorizing Silence

Any attempt at a taxonomy of silence is inevitably something of a simplification. Demystification of Pärt's technical use of silence in this way is, however, useful both as a way of experiencing Pärt's music and of ensuring the correct performance and appreciation of these silences.

Silence is present in the pre-tintinnabuli works. It is not, however, until after 1976 that it becomes a significant and much more intentional compositional procedure. In tintinnabulation it is a fundamental compositional device and is evidenced, for example, in the unusually long pauses in *Missa syllabica*. In the opening of the "Kyrie," each textual phrase is followed by a silence specified by a symbol that indicates the number of quarter-note beats that occur at each double bar (Figure 1).

Missa syllabica demonstrates the principal use of silence in Pärt's tintin-

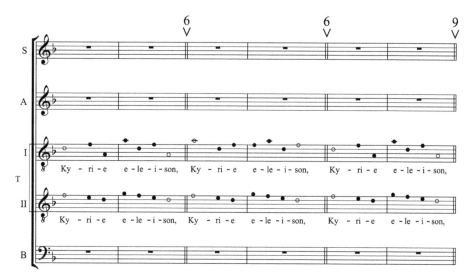

Figure 1. *Missa syllabica*, opening of the "Kyrie." Arvo Pärt "Missa syllabica | für gemischten Chor und Orgel" © Copyright 1980, 1997 by Universal Edition A.G., Wien/UE30431 UE34966.

nabuli (and post-tintinnabuli) style. It is used structurally to shape the form of the movement and delineate phrases; however, these silences are disproportionately long and take on additional meanings and functions that can be deduced from their context and their relationship to the theology of the text. The following list describes the main ways that Pärt uses silence, with examples from his music.

1. Structural Silence

Starting at the macro level, it is clear that silence is used to articulate the structure of works, regardless of their length or complexity. This constructional use is often directly related to Pärt's use of text and is dependent on the manner in which Pärt engages with text as a way of deriving musical material. One of the simplest of these is *Silouan's Song*.[25] The piece is based on a text by St. Silouan the Athonite, a writer who fascinates Pärt and whose work is also the text for *Adam's Lament*. The text for *Silouan's Song* describes the author seeking God in grief and tears.[26] As with *Psalom*, the piece is a comparatively short slow movement (around six minutes). It consists of a number of chorale-like phrases that are not in strict tintinnabulation but carry much of the sound of the stricter works. These phrases are separated by rests or GPs, and Pärt describes the text in the music through an exploitation of tessitura and dynamic range (Figure 2).

Structural silence has two subcategories: surrounding silence (with further

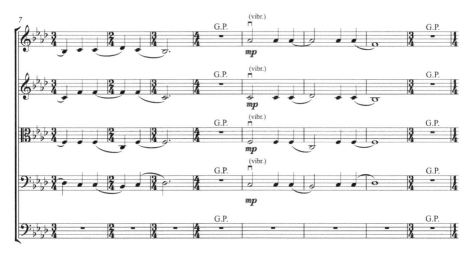

Figure 2. *Silouan's Song*, mm. 7–14. Arvo Pärt "Silouan's Song | 'My soul yearns after the Lord . . .'|für Streichorchester" © Copyright 1991 by Universal Edition A.G., Wien/UE19889.

subcategories of silence before and silence after) and silence that delineates (and is often dictated by) phrase lengths of texts associated, directly or indirectly, with the music. For example, in *Cantate Domino*, the text is sung and therefore clearly part of the compositional procedure, whereas *Für Lennart in memoriam* is an instrumental work that has no sung text but whose musical material is derived from a hymn sung at Orthodox funeral and memorial services.

2. Surrounding Silence

For Pärt, the time before and the time after the music are sacrosanct. As a Christian thinker Pärt is used to the notion of setting aside worldly things and making space for music (an interaction with God). This practice may not be commonplace for his audiences, so the silences that start and end his works are at best a way of quick renunciation, and a preparation for immersion in this fleeting relationship with the Divine, and at worst a useful device to encourage focused and active listening. An element of Pärt's aesthetic is for both the performer and the listener to approach a piece with a reverence that acknowledges the silence before and after.

(a) Emerging from Silence
Pärt eloquently describes the silence and anticipation that precedes music using the analogy of the upbeat:

I imagine the conductor having an upbeat, when the whole thing starts. [This] actually contains the formula of the entire work. Its character, dynamics, tempo, and plenty of other things. . . . I guess the composer is in a similar position before he starts writing.[27]

Although very few pieces start with actual notated silence, several of them emerge from silence in interesting ways.[28] For his setting of the *Te Deum* Pärt speaks in some detail about the compositional procedures that went into the creation of the piece, noting, "I had to draw this music gently out of silence and emptiness."[29] Pärt describes how he abstracted the piece from the ether "by delicately removing one piece — one particle of time — out of the flow of infinity."[30] *Te Deum* begins and ends with a drone (ison) on low D^2. As Laura Dolp observes, "the drone behaves as an aural realization of the silence which precedes and follows the piece."[31] Pärt converts silence into a small sound, focusing it into a single low pitch from which the piece emerges and recedes.

Several other pieces emerge gently from silence. *Lamentate* materializes through the use of the bass drum, *Como cierva sedienta* through a single pitch from the soprano soloist, *Litany* with a triangle stroke and a mysterious chord in the upper strings that starts with a high B^6, and *Es sang vor langen Jahren* has a quasi-cyclical feel that begins and ends with tremolo strings and closes with two measures of rest. *Annum per annum*, for solo organ, makes use of a particular feature of the instrument by starting with the wind supply off. The blowers are switched on as the performer begins playing a series of repeated chords, producing an eerie effect as the pipes begin to speak.

In complete contrast, the arresting opening phrase of *Adam's Lament* surges out of silence and proclaims, "Adam, father of all mankind, in paradise knew the sweetness of the love of God." This strong, startling opening comes as a prelude to the lament that follows, and Pärt's music (all of which is derived from the text) transitions from the power and beauty of Paradise to the agony and sadness of exile.[32]

Other pieces, like *Mein Weg hat Gipfel und Wellentäler* and *Littlemore Tractus*, have a rhythmic steady pattern resulting from precompositional decisions, so what we hear is, in effect, like catching a fragment of a piece that is continuously being played but only becoming aurally present to us for the duration of the performance. From a performance perspective this suggests that these pieces should end without any ritardando in the final measures.

(b) Receding into Silence

In contrast to the beginnings of pieces, Pärt is much more careful and precise about notating silence at the end of his music. This is, in part, a practical move

Figure 3. *Stabat Mater*, m. 513 to end. Arvo Pärt "Stabat Mater | für Sopran, Countertenor (Alt), Tenor, Violine, Viola und Violoncello" © Copyright 1985 by Universal Edition A.G., Wien/ UE33953.

to avoid breaking the mood by movement or applause. Leopold Brauneiss correctly suggests that the silence at the end of the piece may actually be the goal in many cases. Describing *Silentium*, the second movement of *Tabula rasa*, he writes: "The piece finishes as it began: with a single note followed by silence. However, the musical significance of this closing silence is now very different. . . . A precisely composed, resonant silence of a different kind is the goal of the composition *Silentium*."[33] Brauneiss describes the ending of *Silentium* as "sounding silence," and this notion can easily be applied to many other works, including *Stabat Mater* and *Cantus*.[34]

Whatever the motivation for the final silence, it is achieved in several different ways. In many pieces, there is a need for an extended period of the "tonic" mode before the silence. *Es sang vor langen Jahren*, for example, ends with a six-measure A-minor chord and two measures of notated silence. More dramatically, in *Stabat Mater* the coda continues the mood of the piece and includes more than seven measures of the final A-minor chord followed by more than four measures of notated silence (Figure 3).

Many works end with at least a measure of silence or a GP. These include *Partita*, *Quintettino*, Symphony No. 2 (movements 2 and 3), Symphony No. 3 (movement 2), *Tabula rasa* (*Silentium*), *Ein Wallfahrtslied*, *Festina lente*, *Sieben Magnificat-Antiphonen* (movements 1 and 7), *Berliner Messe* (the "Agnus Dei"), *Silouan's Song*, *Kanon pokajanen* (movements 1, 3, 8, and 9), *The Woman with the Alabaster Box*, *Triodion* (all three odes end with GP, but the piece doesn't, since there is an additional doxology), *Zwei beter*, *Orient & Occident*, *Cecilia vergine romana*, *Nunc dimittis*, *Salve Regina*, *Estländler*, *Scala cromatica*, *The Deer's Cry*, *Alleluia-Tropus*, Symphony No. 4 (movements 2 and 3), and *Habitare fratres in unum*. Some of these predate tintinnabulation, so Pärt was aware of the effect this silence had on listeners from early in his career but cultivated it as part of the tintinnabuli style. It is interesting to note that this list includes texted and untexted works and a wide variety of styles, genres, and lengths.

3. Delineation of Phrases

The final subcategory of structural silence is its use as delineation of phrases. This is more sophisticated in many works than the simple division of text noted for *Silouan's Song*. Its most notable early use is in *Passio*, Pärt's setting of the St. John Passion for choir, soloists, and small ensemble. *Passio* was the first large-scale work using strict tintinnabulation principles, and, in order to sustain the seventy minutes of music, Pärt devised elaborate precompositional rules, dictated by the text. These include many used in earlier works, such as setting one word per measure and having words ascend or descend to a central pitch, with the number of pitches equal to the number of syllables in the word. For *Passio*, though, these rules are more complex and include different modes (key areas) for different characters and a sophisticated relationship between text and music that includes notations for punctuation marks. For example, in the last word of a phrase ending with a comma, the stressed syllable is medium, whereas in the last word of a phrase ending with a colon or period, each syllable is long.

What is perhaps most unusual for a work of this length is the number of rests in the music. In total there are 1,533 measures in the piece (of varying lengths, since each measure of text is given to a single word, and its length depends on the number of syllables in the word and the precompositional rules Pärt has applied to them). Of these 228 are complete measures of silence, and a further ninety-seven are measures that contain some silence at the end (which, given the flexible number of beats in a measure of silence amounts to the same thing). This is a remarkable frequency of rests, occurring on average between every five measures of music, which Pärt has used in other works with the same emotional intensity and drama, including his *Te Deum* and *Miserere*.

In *Passio*, each clause is separated by a measure of rest that is usually half the length of the preceding measure but not less than two beats long. Pärt provides more than merely a depiction of the emotional content of the text in the music; he takes the listener on a spiritual journey in which they are effectively invited to a self-emptying in order to gain transcendence. The effect of this amount of silence in a piece that is largely static is extraordinary and a fitting preparation for the concluding triumphant blaze of D major that ends the work.

4. The Silence of Breathing

All musicians breathe in the music they play, even if their sound is not produced directly from the breath. For Pärt, this essential embodiment of silence is important not only at the start of a piece (the inhalation of anticipation) but

Figure 4. Opening of *Miserere*. Arvo Pärt "Miserere | für Soli, gemischten Chor, Ensemble und Orgel" © Copyright 1989 by Universal Edition A.G., Wien/UE30871.

during it, too, and he has made special note on occasion of how this manifests itself. For example, in *Miserere* the work begins with the tenor soloist chanting "miserere mei Deus." Each of his words is followed by a measure of silence, one bar of music for solo clarinet, and another bar of silence. The clarinet forms the first chord of the tintinnabulation, with an arpeggiated E-minor triad (Figure 4).

Pärt described the composition this way: "[*Miserere*] is so structured that for each word there is a catching of breath, a rest, as if someone speaks a word then immediately afterwards tries to gather strength for the next."[35] He describes the text of *Miserere*, which sets the emotionally wrought penitential Psalm 51, as "a chain, in which breathing in and breathing out are interwoven, hope and despair," and he has moved this breathing into the music in a tangible way.[36]

5. Length of Silence

As I noted with *Missa syllabica* and *Passio*, the use of silence is often seemingly disproportionate to the amount of music until one takes into account possible theological motivations. How, then, does Pärt manage to make silence as important as music so that it does not seem as absence or negation but acquires musical properties? It is because silences are not isolated: They are associated with what comes before and after (even the long final silences).

Pärt has not always been successful in the proportion of rests and music. He recalled that the premiere of *Te Deum* was not a success and attributes this in part to the fact that "the audience wasn't used to hearing so many rests, and that made it difficult for them to follow the logic of the work."[37] As a result he made some changes: "After this premiere I revised the lengths of the rests, and I am actually not happy about this, since perhaps something of the original idea has been lost."[38] A taxonomy of the technical way Pärt indicates length of silence is discussed below.

Pärt is concerned with the narrative arc of his music, and this often contains what appear to be disproportionally long silences, either individually or cumulatively. There is a subtle psychology at work here, and usually Pärt is successful in manipulating the sounds to contain silence that can be experienced on many levels.

6. Sounding Silence

Hillier suggests that the way to "play" silence is to link it to the surrounding sounds.[39] In this manner, silence is colored and essentially informs what comes before and after. The absence of pitch is aurally refreshing, and silences often contribute to the calm and spiritual tone of the music (although there are certainly times when these silences can be deafening, pregnant, thick, and heavy). Whether Pärt's silences are soft or loud depends on their position in each piece; however, in the strict tintinnabuli pieces rests are often arranged so that the preceding chord is carried into silence. Like an image burnt on the retina, which can be seen with closed eyes, these tintinnabulation chords become chords of silence burned into the ear and "heard" in the silence. This technique is exemplified in Pärt's extraordinary *Lamentate*, his work for solo piano and orchestra written in response to the sculpture *Marsyas* by Anish Kapoor, exhibited on enormous scale in the Tate Modern Gallery in London. During visits to the space Pärt was aware of the figurative sound of silence in the room and noted, "I have taken the incessant humming of the power plant in the building next door and made it part of my score — the whole tonal scope of the work revolves around this note [A flat]."[40] In conversation with Enzo Restagno he observed that "rests are rich in sound."[41] This means that we have to acknowledge that there may be additional "noise" during a performance and that, even if there is no noticeable sound such as the humming of the power plant, the resonance of the performance space may contribute reverberation to the sound and fill the rests with actual sound rather than the mindful echo of tintinnabuli chords.

7. On the Music between the Silences

If music frames silence and silence is the priority, part of explaining silences is to explain the music. I have done this here, but there are some cases where, according to Pärt himself, some of the music he has written is totally utilitarian. For example, *Stabat Mater* is basically a symmetrical structure with ten stanzas separated by instrumental interludes that are of a different character than the texted sections, being much faster in pacing and with more agitated writing. One of the most important musical elements of *Stabat Mater* is the

Figure 5. *Stabat Mater*, Figure 11. Arvo Pärt "Stabat Mater | für Sopran, Countertenor (Alt), Tenor, Violine, Viola und Violoncello" © Copyright 1985 by Universal Edition A.G., Wien/UE33953.

introduction of these seemingly incongruous fast sections for strings at Figure 11 (Figure 5 above) and Figure 18.

Each lasts only a few measures (twenty-two and twelve, respectively), but they have been criticized for disrupting the quiet introspection of the rest of the piece. Viewed holistically it is easy to explain these on several levels: The instrumental materials are clearly derived from the sung material; the material is closely related between instruments; they act as a kind of aural sorbet, cleansing the listening palate before the next episode; and they are evocative, reflecting the emotion of grieving, which comes in waves of experience. Describing these interludes (comparing them to the similar ones in *Miserere*) Pärt says, "These are just—I'm not sure quite how to put it—just music: music we need, like light, like air."[42] This is a description of a purely functional music whose principal purpose is part of the aesthetic and which Pärt intuits (or designs).

Notation and Performance

Given the comparatively small number of signs available to denote silence in a musical score, being specific is quite difficult. Pärt's scores are sometimes hard to realize because his notation is ambiguous and inconsistent over his career. This is not to be understood as a criticism, merely an observation; however, performers have to interpret these ambiguities. Rather than rely on the vagueness of intuition or "musicality" (both largely a product of a kind of formative training that does not necessarily help with a stylistically appropriate performance of Pärt's music), we can begin to understand how to perform these marks, taking into consideration their possible theological function (for example, the chain breathing vacillating between hope and despair in *Miserere*) and the applicability of the taxonomy just described.[43] In particular, performers need to think through the issues of notation and intention so that an overreliance on replicating recordings (especially authoritative recordings that have the composer's imprimatur) does not result in a standardization of performance.

In order to identify the issues raised by specific signs in Pärt's scores, it is useful to deal with them individually as they relate to specific instances.

Rests and New Notation

The most common form for indicating silence is the rest, and Pärt uses rests liberally and precisely. If there is any ambiguity, the value of the rests is usually notated with a numeral above the measure at the top of the score (as in *Passio*, for example). With some pieces, such as *An den Wassern zu Babel saßen wir und weinten* and *Missa syllabica*, Pärt experimented with new forms of notation for both music and rests. In *Missa syllabica* the note heads do not have stems but are essentially the same as the equivalent normal value with stems. Pärt only uses the quarter-, half-, and dotted half-note values and therefore does not encounter problems of having to find a different notation for any value less than a quarter note. There is a new notation for the number of beats of rest between sections (a number above a large V inserted over a double bar line at the end of each phrase of the text). In the version for choir and string quartet from 2009, Pärt adds an indication for the length of the comma because of apparent confusion (Figure 6).

Lunga Pausa

The longest of the traditional notations for silence is not a frequent sign in Pärt's music, perhaps because of its inherent ambiguity. Pärt uses it at the end of *Veni creator*, where "molto lunga" is marked over the final chord. As a result of questions from performers, the score indicates that the duration of the pause is roughly equivalent to eight measures. No silence following the chord is indicated, though it is obviously expected. By contrast, in *The Deer's Cry*, Pärt has notated almost sixteen beats of rest, ensuring the piece ends by returning to the silence from which it came. Here there is a clear expectation that the rests are part of the piece and should be counted out before any postpiece silence (even if the transition is not apparent to the listener).

Grand Pause

It is possible that the GP for Pärt means something slightly different than a notated measure of silence. The difference might be that the GP implies some-

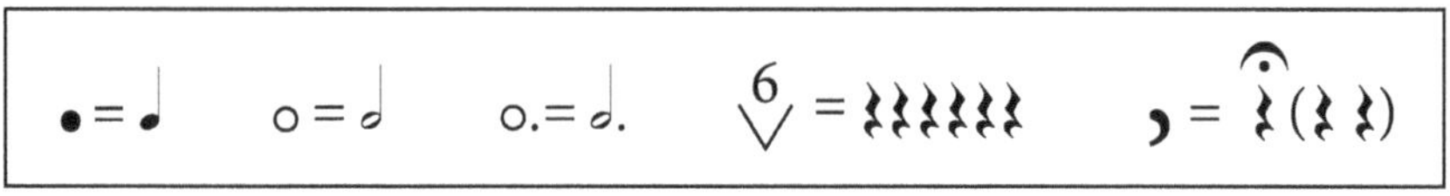

Figure 6. *Missa syllabica* notation (2009 version). Arvo Pärt "Missa syllabica | für gemischten Chor und Orgel" © Copyright 1980, 1997 by Universal Edition A.G., Wien/UE30431 UE34966.

thing less precisely notated and more reliant on the mood of the performance and the intuition of the performers than the notated version (which could be counted out precisely). For example, in the second printed version of *Stabat Mater*, the last four measures of the piece each include the notation GP (see Figure 5, from the earlier score, which does not have the additional markings). Because Pärt made this notational change, we know that he wanted to modify the silence in some way; providing the GP notation might imply a freedom to exploit the "sounding silence" that he did not feel was afforded by the stricter notation. It is certain that Pärt expects performers to include and account for the resonance and reverb of the performing space in calculating the length of any notated rests.

The Fermata

In *Psalom* each of the ten phrases of unheard text is separated by a GP (though, perhaps surprisingly, there is not one at the end of the piece, which ends with a fermata over the last chord). These GPs are of varying lengths of either nine or twelve beats (there is no tempo indication), with the exception of the last GP (m. 72), which is notated the same as the others but with a fermata rather than a numerical value for the rest. It is possible that this is connected with theories of rhetoric in which, for this last fermata, the conductor is empowered to respond to the progress of the performance and to adapt the pause so that the final phrase is the most telling. This "emphatic pause" requires some skill to communicate to the audience, since *Psalom* is a work for strings that is based on Psalm 112 (113), and unless there are program notes to explain the mechanics of the work, audiences will not be aware of the textual correspondence of each musical line and the emphasis Pärt places on the final phrase "Praise ye the Lord" (Figure 7).

Bar and Double Bar Lines

In stricter tintinnabuli works Pärt separates each word with a bar line. These do not indicate any break in the music, which are generally indicated by double bar lines. There is experimentation in some pieces, such as *And One of the Pharisees . . .* for unaccompanied ATB choir. There are new note values,

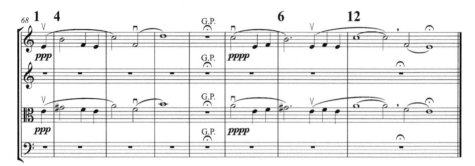

Figure 7. *Psalom*, m. 68 to end. Arvo Pärt "Psalom | für Streichorchester" © Copyright 1991, 1997 by Universal Edition A.G., Wien/UE30847.

which, unlike *Missa syllabica*, only hold some correspondence with normal notation. Pärt also includes two bar line notations that are explained in the score. The longer pause comprises two lines, the first of which is dotted. The shorter pause is just a dotted bar line. In Pärt's explanation the length of rests is not precise but indicated by a larger fermata for the longer pause and smaller one for the shorter silence (Figure 8). It is left to the conductor to decide on the proportion of the relationship between the music and the silence.

Comma and Multiple Signs

In *Magnificat* words are delineated by dotted bar lines, and clauses are delineated by double bar lines. At the end of most phrases the bar lines are the only marking; however, at the end of measure 6 ("qui potens est") there is an additional comma in the score. Four further commas are marked, but they are not consistent as to their placement within the text (for example, they do not denote a period or other pause), and they are not at any particularly strategic points in the musical narrative. Recordings by both Cleobury and Hillier adopt the tempo indication noted in the score, but Cleobury does not acknowledge any difference in the phrases that end with the additional comma except where there is a final consonant that occupies an extra "beat." Cleobury also inserts many more breaks for breath mid phrase and gives them effectively the same space as the ends of phrases indicated by double bar lines. Although this sounds natural, it does not conform to an executor's faithful interpretation of the score but rather one of an interpreter.[44]

Beatus Petronius is a tintinnabuli work for two choirs and two organs. This score has performance issues, since Pärt indicates silence with pause marks and parenthetical whole-note notations over a bar line (see measure 14) in the two revised versions. Sometimes, however, he also adds a comma (see measure 4), and it is not immediately clear what mathematical length this might

Figure 8. *And One of the Pharisees . . .* , notation. Arvo Pärt "And One of the Pharisees . . . | für 3 Singstimmen oder 3-stimmigen Chor a cappella" © Copyright 1992 by Universal Edition A.G., Wien/UE30510.

Figure 9. *Beatus Petronius* (revised, 1996), mm. 4–5. Arvo Pärt "Beatus Petronius | für 2 gemischte Chöre (SATB/SATB) und 2 Orgeln" © Copyright 1990, 1997 by Universal Edition A.G., Wien/UE31156.

have, since there are three indications here, two of which are vague and one of which is precise (Figure 9).

One of the most important questions for Pärt's music is whether the silences need to be mathematical. In fact, what is generally required is a silence of no fixed duration but long enough for quiet to fall in the acoustic and for the space between the sounds to contain a silence that can be felt. On the other hand, in a dead acoustic some of the long silences Pärt notates in *Missa syllabica*, for example, can feel artificial, even uncomfortable, and could be shortened at the discretion of the conductor.

What is important to note is that for performers these rests are not merely for counting; they are substantive and require an attitude toward performance that is perhaps different from normal.

Editing Pärt's Works

Since it is clear that Pärt's use of rests is an attempt to notate something that is not notatable (to "express the inexpressible," as Huxley puts it), is the choice of imprecise notation precisely calculated, or do Pärt's works need to be edited? Performers have been confounded by some of the marks, and this has led to certain revisions by Pärt himself, some of which have been described here. Pärt commonly works on pieces in rehearsal and during recording sessions, refining and crafting as necessary. In some of the later works he has acceded to the requests from performers for more guidance regarding issues such as tempo and dynamics.

An extreme version of this is the score for *Te Deum*. The piece was premiered and recorded by the Estonian conductor Tõnu Kaljuste, who suggested a detailed plan of markings generally absent from Pärt's scores. The preface to the score includes an interesting disclaimer: "The metronome details in this score are from the conductor Tõnu Kaljuste for [the] ECM CD production and are not binding."[45] From this we may infer that Pärt was satisfied with Kaljuste's version but can envisage other satisfactory solutions. It is evident from various comments he has made that this variety of approach can also be applied to the performance of his rests.

Although silence is a complex issue for Pärt's music, overlaid with extra-musical meaning, it is clear that it is always directed to the goal of transcendence. In Pärt's soundworld, where words and music fail, silence expresses the ineffable, and that is always and everywhere Pärt's goal.

Notes

1. This chapter's epigraph is from Aldous Huxley, "The Rest Is Silence," in *Music at Night and Other Essays* (London: Chatto and Windus, 1931), 22.

2. *Current Biography Yearbook*, 56th ed. (New York: H. Wilson Co., 1995), 456.

3. Paul Hillier, *Arvo Pärt* (Oxford: Oxford University Press, 1997), esp. 7–10; Andreas Peer Kähler, "Radiating from Silence: The Works of Arvo Pärt Seen Through a Musician's Eyes," in *The Cambridge Companion to Arvo Pärt*, ed. Andrew Shenton (Cambridge: Cambridge University Press, 2012), 193–97.

4. Leo Normet, "The Beginning Is Silence," *Teater. Muusika. Kino* 7 (1988): 19–31.

5. Adriana Helbig, "Arvo Pärt: The Search for the Eternal Silence at the Heart of Sound," BA thesis, Drew University, 1997. For a discussion of the spiritual and ontological status of silence in *Sarah Was Ninety Years Old*, see Marcel Cobussen, *Thresholds: Rethinking Spirituality Through Music* (Aldershot: Ashgate, 2008), 107–24; Mark Vuorinen, "Symbolic Chiasm in Arvo Pärt's *Passio* (1982)," *Circuit* 21, no. 1 (2011): 45–59.

6. Peter C. Bouteneff, *Arvo Pärt: Out of Silence* (Yonkers, NY: SVS, 2015).

7. Bouteneff, *Arvo Pärt*. These pervade the text but are encapsulated on 103. See also 109, 117, 121, 125.

8. Bouteneff, *Arvo Pärt*, throughout chap. 2, and summarized on 134.

9. Arvo Pärt, interview with Tom Huizenga, in "The Silence and Awe of Arvo Pärt," NPR, June 2, 2014: https://www.npr.org/templates/transcript/transcript.php?storyId=316322238.

10. Bouteneff, *Arvo Pärt*, 129.

11. See, for example, Allan Ballinger, "In Quest of the Sacred: Arvo Pärt and *Sieben Magnificat-Antiphonen*," DMA diss., University of Connecticut, 2013; Maria Cizmic, "Transcending the Icon: Spirituality and Postmodernism in Arvo Pärt's *Tabula Rasa* and *Spiegel im Spiegel*," *Twentieth-Century Music* 5, no. 1 (March 1, 2008): 45–78; Laura Dolp, "The Silent Garden: Inventive Procedure and Ancient Principle in the *Berlin Mass* of Arvo Pärt," MA thesis, Boston University, 1997; and Hillier, *Arvo Pärt*, 8ff.

12. Hillier, *Arvo Pärt*, 8.

13. Pärt, interview with Huizenga.

14. Normet, "The Beginning Is Silence," 22.

15. For my more detailed discussion of Pärt's theology, see "Interlude: The Numinous Encounter," in *Arvo Pärt's Resonant Texts: Choral and Organ Music, 1956–2015* (Cambridge: Cambridge University Press, 2018), 88–104.

16. Susan Sontag, "The Aesthetics of Silence," in *Styles of Radical Will* (New York: Farrar, Straus and Giroux, 1966), 3.

17. Sontag, "The Aesthetics of Silence," 3.

18. Sontag, "The Aesthetics of Silence," 4.

19. Sontag, "The Aesthetics of Silence," 4, 5.

20. Sontag, "The Aesthetics of Silence," 11.

21. Sontag, "The Aesthetics of Silence," 13.

22. Sontag, "The Aesthetics of Silence," 13.

23. Sontag, "The Aesthetics of Silence," 14.

24. "I am tempted," Pärt said, "only when I experience something unknown, something new and meaningful for me. It seems, however, that this unknown ter-

ritory is sooner reached by way of reduction than by growing complexity." Quoted in Geoffrey J. Smith, "Sources of Invention: An Interview with Arvo Pärt," *Musical Times* 140, no. 1868 (Fall 1999): 19.

25. For a discussion of Pärt's pieces for instrumental ensemble with silent texts, see Toomas Siitan's chapter in this volume.

26. For the full text, see Hedi Rosma et al., eds., *In Principio: The Word in Arvo Pärt's Music* (Estonia: Arvo Pärt Centre, 2015), 335.

27. Transcribed from Dorian Supin's film *Arvo Pärt: 24 Preludes for a Fugue* (2002).

28. A notable exception is *Cantus in Memory of Benjamin Britten* (1977), which begins with notated silence and the regular tolling of a bell, interspersed with silence, before the mensuration canon that comprises the body of the work begins.

29. CD liner notes, *Te Deum* (ECM 1505NS, 1993).

30. Hillier, *Arvo Pärt*, 140.

31. Dolp, "The Silent Garden," 81.

32. "The content and structure of the texts . . . dictate the course of the music down to the smallest detail. Punctuation, syllable counts and word emphases all play decisive roles in the composition." https://www.universaledition.com/sheet-music -shop/adam-s-lament-for-mixed-choir-satb-and-string-orchestra-paert-arvo-ue34740.

33. Leopold Brauneiss, "Tintinnabuli: An Introduction," in *Arvo Pärt in Conversation*, ed. Enzo Restagno et al., trans. Robert Crow (Champaign, IL: Dalkey Archive, 2012), 114.

34. Leopold Brauneiss, "Musical Archetypes: The Basic Elements of the Tintinnabuli Style," in *The Cambridge Companion to Arvo Pärt*, ed. Andrew Shenton (Cambridge: Cambridge University Press, 2012), 49–75.

35. Enzo Restagno, "Arvo Pärt in Conversation," in *Arvo Pärt in Conversation*, ed. Enzo Restagno et al., trans. Robert Crow (Champaign: Dalkey Archive, 2012), 67.

36. Restagno, "Arvo Pärt in Conversation," 69.

37. Restagno, "Arvo Pärt in Conversation," 49.

38. Restagno, "Arvo Pärt in Conversation," 49.

39. Hillier, *Arvo Pärt*, 199.

40. https://www.universaledition.com/composers-and-works/arvo-part-534/works /lamentate-10864.

41. Restagno, "Arvo Pärt in Conversation," 68.

42. Restagno, "Arvo Pärt in Conversation," 74.

43. See my essay for an evaluation of the types of training useful for stylish Pärt performance: "Performing Pärt," in *White Light: Arvo Pärt in Media, Culture, and Politics*, ed. Laura Dolp (Cambridge: Cambridge University Press, 2017), 212–37.

44. *Ikos: Choral Music by Górecki, Tavener, Pärt*, Choir of King's College Cambridge/Cleobury (EMI Classics, 1995); *De profundis*, Theater of Voices/Hillier (Harmonia Mundi, 1997). For more on the distinction between an executor and an interpreter, see Shenton, "Performing Pärt," 220–23.

45. Preface to the 2009 score.

IV

Materiality and Phenomenology

8
Vibrating, and Silent
Listening to the Material Acoustics of Tintinnabulation

Jeffers Engelhardt

And you find, if all the elements of a chord are absolutely in tune, then the harmony is what we call "locked in," or "in tune." And suddenly, in what seems to be a very small, simple chord . . . the overtones start, it blossoms, it becomes a much larger sound, you get these ringing tones, the harmonies start to resonate. Arvo never writes a note or changes it unless it means something, and that's why I admire him so greatly.

— DAVID JAMES, THE HILLIARD ENSEMBLE

There has to be, however, a balance between the human perceptive faculty and the musical presentation. All important things in life are simple. Just look, for example, at the partial tones of the overtone scale: the initial, lower overtones are perceptible and easily distinguishable, whereas the upper ones are more clearly defined in theory than audible.

— ARVO PÄRT

Pedal Down

The first notes of *Für Alina*, low Bs on the piano, B0 and B2.[1] (The octave registers of the piano are as shown in Figure 1.) Pedal down, dampers up to create acoustic space and possibility—a cathedral in the piano. Pedal down, freeing those Bs, those strings, that material, to vibrate at their fundamental frequencies—around 31 Hz and 123 Hz—and to vibrate through a spectrum of higher partials[2] (roughly the octave, B; a fifth above that octave, F#; two

129

Figure 1. The octave registers of the piano.

Für Alina
für Klavier (1976)

Arvo Pärt
(*1935)

Ruhig, erhaben, in sich hineinhorchend

Figure 2. *Für Alina*, mm. 1–3. Arvo Pärt "Für Alina | für Klavier" © Copyright 1990 by Universal Edition A.G., Wien/UE19823.

octaves above the fundamental, again B; almost, but not quite, a major third above that, D#; another fifth, F#; a seventh, thirty cents[3] flat, A; and so on). Pedal down, so that the vibrating energy from those two Bs sets other material in motion, other courses of strings vibrating sympathetically. Free to resonate, more Bs, F#s, D#s, As, and higher partials start to sound, adding another dimension to the sounding of those low Bs — the sympathetic resonance that materializes the extent of the piano's "sonic body"[4] and, in a line from antiquity through nineteenth-century scientist/philosophers like Joseph Fourier and Hermann von Helmholtz to digital signal processing, connects a listener to the material and psychoacoustic properties of sound.[5]

Half a second into *Für Alina*, after the transient (the high-amplitude, noisy onset of hammers hitting strings) fades enough, combinations of fundamentals and partials materialize and kaleidoscope through the next several seconds of resonant decay. (In Pärt's score, shown in Figure 2, those low Bs are suggestively tied to a staff that dissolves into plain white paper.) Sitting at the piano, listening intently to those several seconds of resonance (see the spectrogram in Figure 3 and play Audio Example 1, available at https://youtu.be/2kgHhkcNTzg) before moving my hands up to B4 and C#6, I hear the gentle pulsing of acoustic beats[6] and shimmering, metamorphosing timbres as a spectrum of frequencies is activated up to and off of the right-hand edge of the piano keyboard, audible to many ears but no longer playable on the instrument. (Don't let anyone tell you that C8, the last key on the keyboard, is the highest sound on the piano at around 4188 Hz.) Out of the resonant decay of those low Bs,[7] I hear D#3 (around 156 Hz), F#4 (around 370 Hz), and a flat A5 (around 865 Hz) with particular clarity, rising and falling in the mix of other resonances and oscillating beats.

Pedal down, those low Bs make the beginning of *Für Alina* (and the rest of the piece, and lots of Pärt, as I will argue) "about" the psychoacoustics of intent listening and a listening subject's embodied relationship to vibrating material.[8] Here, I write like a pianist listening to their instrument in the "Solitudine–stato d'animo" section of *Lamentate* to make sure their articulation, pedaling, and voicing are attuned to its resonance and pulsating timbres (see the Cizmic and Helbig chapter in this volume); or like a singer attenuating their vibrato, shaping the vowels, and tuning tight intervals according to the rate of acoustic beats at the climax of *Nunc dimittis* on the word "Israel"; or like the Goeyvaerts Trio performing *Stabat Mater* in just intonation and without vibrato[9] to better realize the "voice" of music in which "the wave frequencies of the individual notes are not always 'pure'";[10] or like a recording engineer working in a tried-and-true space for Pärt recordings, like the Church of St. Nicholas in Tallinn, to capture the sonorousness of Pärt's music as "intermaterial vibration."[11] In each of these cases, there are possible (but not, of course, universal) modes of listening available to those who seek out the acoustic phenomena of Pärt's music — ways of listening to music not only for its meanings (what it figures, symbolically represents, or enables through performance), its qualities (the values and identities a sound mediates), or its referential capacities (structure, tuning systems, genre) but for its "verberation,"[12] its existence as vibrating material.

Rehearing Silence

These possible modes of listening are "silent" in the sense that music, as vibrating material, offers no message or code. Listening subjects "silence" music as energy and medium, haptic and resonant perception. In Pärt's work, such silence (in contrast to Shenton's meditation on silence in this volume) compels us, silences us, to listen to acoustic phenomena per se — the resonant frequencies of spaces, the timbral shimmer of voices, or the inharmonicity of a

Figure 3. Spectrogram of the low Bs at the opening of *Für Alina*.

piano — not for what they mean or index but, in a rarified way, for what they are as phenomena. Pierre Schaeffer, in his foundational *Traité des objets musicaux* (1966), terms this "reduced listening"—a possible, "Schaefferian" approach to listening that I entertain throughout my writing here.[13] For Schaeffer and the strains of sound studies his work inspires, reduced listening "put[s] listening on the rack between this event and that meaning."[14] By "ceasing to listen to an event through the intermediary of sound," Schaeffer writes, "we will still be listening to the sound as a sound event."[15] Michel Chion takes this up, elaborating on reduced listening as listening to, not listening through; listening to the sounds of language, not to the text; listening for "those sensible qualities not only of pitch and rhythm but also of grain, matter, form, mass, and volume"; listening to sound "as an object of observation in itself, instead of cutting across it with the aim of getting at something else."[16] In the silence of reduced listening, one listens not for locations, identifications, or causes like humans are evolutionarily predisposed to do but to acoustic phenomena as something to behold. Pärt's silence, emerging through his idiomatic approach to the materiality of instruments and voices and the psychoacoustics of listening, can be a mode of attunement to things as they are perceptually.[17] In this way, attending to silence is an ethical practice of listening that puts in parentheses (Husserl's *epoché*) the process of perception, turning it back on itself. This is shaped, of course, by the social and cultural positions of listening subjects and what they desire from or respond to in Pärt's music, and it reorients around sound and embodiment discourses on the algorithms and forms of "tintinnabuli," Pärt's musical engagement with word and language, and the cultural and spiritual dynamics of his work.

In practice, reduced listening relies on an acousmatic engagement with sound — with "sound unseen";[18] the mythical Pythagorean voice behind the veil; the sound per se, perceptible by "turning [one's] back on the instrument and musical conditioning"[19] that is, nevertheless, "diabolically difficult" to behold.[20] Mediation plays a crucial role in the possibility of acousmatic listening, decoupling a sound from its cause, embodiment, and audiovisual complex. The voiceover in film, heartbeats and breathing heard through a stethoscope, audio recordings of music, broadcast sound heard from afar, or an individual voice in a singing chorus or chanting crowd — each of these is acousmatic. And when music silences us into a mode of reduced listening to the phenomena of its material vibration, when one works at listening to the acoustic beating, shimmering timbres, and kaleidoscoping partials of *Für Alina's* low Bs, this, too, is acousmatic experience, even when the piano and the pianist are right there.[21]

The silence of acousmatic listening that coincides with Pärt's material acoustics is the silence of nonrepresentational experiences of sound and music.[22] It is not the cessation of vibration but one of the ways sound is experienced not for what it means, communicates, or signals (in this sense, acousmatic listening is as "unnatural" as other modes of listening). It is not music representing silence by an aesthetic framing of music's other (cf. Shenton in this volume) — it is not music stopping, pausing, or fading out to evoke metaphorical silence.[23] It is also not the Cagean reconceptualization of listening to the metaphorical silence of ambient sound as meaningful and anything but silent.[24] It is not the indexical silence of the Japanese concept of *ma*, the "relationship between human perception and the natural environment, characterizing the 'in-between-ness' of space and time inherent in the audition of sound."[25] Nor is it the silence of silent reading and prayer or audiated music, which, although not involving vibration, involves the sounds, memories, and imaginings of speech and music that mediate things beyond themselves.[26] To reiterate: The silence I write about here is the energy of vibrating material and its perception as nonrepresentational, nonfigurative sound. Sometimes this silence is fleeting, lasting just a few seconds until sound becomes a sign, takes on meaning, or elicits emotion.[27] And sometimes this silence is longer, lasting for the duration of the music that silences through its acoustic phenomena. In human terms, though, all silence is metaphorical, anthropocentric, and a product of human limitations — silence as what is below or beyond thresholds of hearing and haptic perception; silence as the absence of a medium, including a body, for vibration; or silence as the absence of humans, either in the tree-falling-in-a-forest thought experiment or in the awesomeness of death and threats to our survival as a species in the Anthropocene.

Rethinking Silence

The acoustics of silence are baked into the fundamental principles of Pärt's tintinnabuli — pitches in the major triad materialize as the most strongly present harmonics; uncomplicated, homorhythmic textures, slow tempi, widely spaced intervals, and long note values work together to lock vibrations into relationships of interference and resonance; and the ubiquity of dissonant tintinnabuli clusters (lots of seconds, sevenths, and ninths) guarantees that shimmering, pulsating, beating timbres infuse the Pärt sound.[28] Performers and listeners are attuned to this, which is why dispassionate, minimal-vibrato, flexibly tuned choral and instrumental performances are what seem right. Attending to these material, acoustic dimensions, then, is the sonic turn in thinking about Pärt whose moment has arrived.

Silence — always metaphorical, including the sense in which I use it here—is one of the master tropes in conversations about Pärt. Pärt himself circles back again and again to different notions of silence in his life and music. In the segment of the fourteen-part radio documentary *Arvo Pärt 70* (2005) produced by Immo Mihkelson titled "Silence," Pärt describes his discovery of tintinnabuli as intimately linked with his discovery of inner silence (*vaikus*), a silence that he contrasts to the inner noise (*müra*) he once needed to write pre-tintinnabuli music. What is that silence? Apparently reading from his compositional diaries from the 1970s, Pärt offers up customarily aphoristic, ersatz explanations infused with Orthodox Christian language: "Silence is wisdom. Silence—you must approach it with great awe. Silence judges us, not just each word and sound, but every thought that might be a sin, unclean."[29] This characterization of silence, resonating within soul and sound, carries over into popular representations and scholarly approaches to Pärt. The canonical quotation, first appearing in Wolfgang Sander's liner notes to the 1984 ECM *Tabula rasa* album that brought Pärt to global attention and has circulated in countless contexts thereafter, reads like this:

> Tintinnabulation is an area I sometimes wander into when I am searching for answers—in my life, my music, my work. In my dark hours, I have the certain feeling that everything outside this one thing has no meaning. The complex and many-faceted only confuse me, and I must search for unity. What is it, this one thing, and how do I find my way to it? Traces of this perfect thing appear in many guises— and everything that is unimportant falls away. Tintinnabulation is like this. Here I am alone with silence. I have discovered that it is enough when a single note is beautifully played. This one note, or a silent beat, or a moment of silence, comfort me. I work with very few elements—with one voice, with two voices. I build with the most primitive materials—with the triad, with one specific tonality. The three notes of a triad are like bells. And that is why I called it tintinnabulation.[30]

Some of the earliest public responses to tintinnabuli in Estonia and beyond pick up on the trope of silence. In a segment of the *Arvo Pärt 70* radio documentary titled "The Silence That Changed the World—A New Sound-Language," numerous people involved in the first performance of *Tabula rasa* in 1977 (the conductor Eri Klas; Nora and Arvo Pärt; the musicologists Maia Lilje, Toomas Siitan, and Merike Vaitmaa; the pianist and composer Rein Rannap; the violinist and conductor Andres Mustonen; and the composer Erkki-Sven Tüür, among others) describe the unforgettable, profound silence that followed the four measures of rest at the end of the "Silentium" movement. Listeners dared not move or breathe, Pärt worried that his racing heart-beat was audible to those sitting around him, and, according to those who were there, several audience members fainted.[31] From this legendary moment, which was reported on in the official Soviet press,[32] and the buzz generated by the ECM *Tabula rasa* album flowed other associations of Pärt's sound and the trope of silence. A 1988 issue of the Estonian cultural magazine *Teater. Muusika. Kino.*[33] featured three articles on Pärt[34] (a sign of his perestroika-era rehabilitation in Estonia) that articulated what are now common invocations of metaphorical silence in his life and music — spiritual, creative, sonic, and as a mode of listening.

From the first sentence onward, silence plays an important role in Paul Hillier's definitive book on Pärt's tintinnabuli work through the early 1990s.[35] Hillier is quick to establish the metaphorical nature of silence in relation to Pärt, reminding readers that, although absolute silence does not exist in human terms, its place in describing Pärt's period of public inactivity in the late 1960s and 1970s, the hesychast tradition of Christian mysticism, and the Enlightenment dialectic of absence and presence figure prominently in how people make meaning around his music and persona. Along with a steady stream of laconic statements from Pärt on silence in documentary films, published interviews, and public speeches,[36] the trope of silence has a durable presence in popular and scholarly conversations about Pärt. Laura Dolp,[37] Kevin Karnes,[38] Andrew Shenton,[39] Toomas Siitan,[40] and others use silence, often in conversation with Arvo and Nora Pärt's own words, to get at the important critical and aesthetic questions that tintinnabuli and Pärt's persona raise. Peter Bouteneff's intervention in the conversation about Pärt and silence elaborates on the substantial connections between silence as a metaphor and the idea of silence in Orthodox Christian theology, Christian personhood, and contemporary spirituality and public religion. "Silence is multivalent," Bouteneff writes — it "is more than just an absence; it is truly a kind of presence. It is a 'thing' or state to be pursued. Or, in the presence of greatness, of awe, we are 'brought' to silence."[41]

The commonplace of Pärt and silence I outline above is not the kind of silence I write about here. Silence that makes something else perceptible, indexes conditions of possibility, or represents something other than itself is not the silence of a Schaefferian mode of listening to vibrating material per se. Pärt's material acoustics — vibrations given musical space, time, and relationship for their intonations, resonances, and interferences to be realized psychoacoustically — can silence the languaging, imaging, and emoting that accompany other ways of listening.[42] This experience of silence is not exclusive to Pärt. The silence of material vibration when it is "about" the auditory and haptic perception of propagating sonic energy is a powerful part of much musical experience — the sound systems of Jamaican reggae,[43] bass-heavy dance music and sound art,[44] or the work of sound artists like Ryoji Ikeda, for instance. In terms of Pärt, then, there is another kind of silence (contrasting to the commonplace of Pärt and silence) that matters: the immersive, sensate, palliative experience of a sonic body touched by vibration.

Tintinnabuli's Silence

In light of this thinking about silence, there are fruitful ways of revisiting the allusions to bells and habits of embodied listening that circulate around tintinnabuli. The etymological and figurative connections between tintinnabuli and the acoustics of bells are well documented.[45] It is also established that Arvo and Nora Pärt's coining of the term "tintinnabuli" in connection with the eponymous 1976 concert at which Pärt's first tintinnabuli works were debuted was an onomatopoeic evocation and generalized reference to the sonority of the triad, not a description of the algorithmic processes of the technique.[46] So while tintinnabuli "merely alludes to the sound of a bell,"[47] the sounding of tintinnabuli — its qualities as vibrating material rather than notes on a page[48] — can hail listeners, and silence them, through its bell-like touches on the sonic body. As an attitude of listening, in other words, tintinnabulation (in an active sense of Christopher Small's "musicking")[49] has much in common with the acoustic phenomena of ringing bells. In Pärt's characterization, tintinnabulation happens in the material and psychoacoustic workings of vibration; it "means the interaction of two simultaneously sounding musical lines — their sonic relationships."[50]

Looking at the spectrogram in Figure 4 of the tubular bell A4 that begins *Cantus in Memory of Benjamin Britten*, one sees a rich spectrum of partials that produces the psychoacoustic effect of the strike tone — A4 (440 Hz), in this case, which is not present in the frequency spectrum but is, nevertheless, perceived as the fundamental frequency of the bell because of how other partials are tuned nearly harmonically (play Audio Example 2, available at https://youtu.be/ycnyYQs4JoQ). The formants[51] at around A5 (880 Hz), E6 (1318 Hz), A6 (1760 Hz), and sharp C#7 (2310 Hz) are the second, third, fourth, and fifth partials of the "missing" A4 fundamental (the strike tone), which the ear tends to posit as the "pitch" of the tubular bell (although the more intently one listens, the less true this may seem).[52]

Listening intently to the bell at the beginning of *Cantus*, other acoustic phenomena materialize as well, drawing one ever closer toward the sound per se. The formants cover a broad enough range of frequencies that they produce subtle interferences audible as acoustic beating oscillating at a frequency

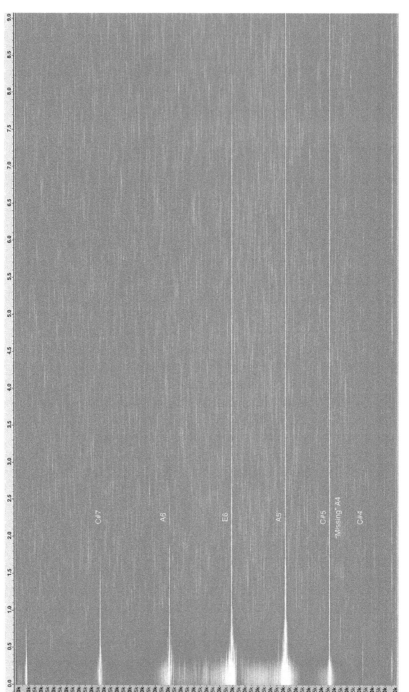

Figure 4. Spectrogram of the A4 tubular bell.

of around 1.6 Hz. What gives this tubular bell (without end plug) its idiomatic sound are the strong formants at C#5 (around 554 Hz) and C#4 (around 277 Hz), which are not part of the harmonic series that produces the virtual fundamental A4. Over the duration of the bell's sounding envelope from transient strike to decay, those C#s shine through the frequency spectrum as higher partials die off, with the effect of the bell "speaking" a vowel sound modulating from open to closed before falling beneath the threshold of audibility. Later on in *Cantus*, those strong C# formants rub up and shimmer against the natural Cs spread across the dense canonical texture of the A-minor music.

The material acoustics of bells and the attitude of listening they summon up capture the engaged sonic embodiment of tintinnabulation, something Hillier adumbrated two decades ago: "If a single bell is struck, and we contemplate the nature of its sound—the *Klang* at impact, the spread of sound after this initial gesture, and then the lingering cloud of resonance—what we hear takes us to the heart of tintinnabuli."[53] There are also intimations of this kind of intent listening to the material acoustics of tintinnabuli coming from Arvo and Nora Pärt. In the "Auftakt" scene of Dorian Supin's documentary *24 Preludes for a Fugue* (2002), Pärt famously describes the kind of intent listening that *Für Alina* requires—a listening "where one concentrates on each sound so that every blade of grass would be as important as a flower . . . in order to see something more than just this white key and this black key in a small sound, a small phrase."[54] In the program notes to the landmark 1976 concert where Pärt's first tintinnabuli works debuted, Nora Pärt, quoting from Arvo Pärt's compositional diaries, offers us this: "What is the significance of a single sound or word? Those thousands that have streamed by our ears have made our receptive apparatus dull. One must treat every sound, word, and act with care."[55] Finally, Nora Pärt hints at a Schaefferian mode of reduced listening to tintinnabuli:

> The concept of tintinnabuli was born from a deeply rooted desire for an extremely reduced sound world which could not be measured, as it were, in kilometres, or even metres, but only in millimetres. According to my experience, the listener becomes increasingly sensitised in the process once he is drawn into this dimension. By the end the listening attention is utterly focused.[56]

A less figurative connection between tintinnabuli and the sounding of a bell, I believe, is in the ways its ringing invites an intent, reduced listening to a shimmering, harmonically rich spectrum of sound. This is a listening "in millimetres" that treats every sound "with care" "so that every blade of grass would be as important as a flower." The sounding of a bell is an ideal form of tintinnabulation, the active outcome of tintinnabuli's algorithms and forms taken up by the sonic bodies of listeners. Tintinnabuli's silence, then, is listening reduced to the phenomena of material vibration; listening to the silence of sonic relationships in the psychoacoustic event in and of itself, not the non-silence of sounds extending beyond themselves. This take on tintinnabuli puts its etymology and evocative quality at the center of an intent listening not just to bells but to all vibrating material.

String Theory

I left off with *Für Alina*'s opening low Bs sounding through beating, meta-morphosing combinations of fundamentals and partials. The lower partials of those B strings — Bs, F#s, and D#s — rise and fall in the audible mix, a crystal-clear demonstration of the immanence of the triad in tintinnabuli and the liquid, kaleidoscopic timbres produced through the subtle inharmonic vibrations of the piano (see Figure 3).[57] The energy from the low Bs and partials, which we can visualize vibrating along a Cartesian z-axis down the length of the piano (see Figure 5), sets other material in motion, sympathetically. Pedal down, the strings of other Bs, F#s, D#s, As, and higher near-harmonics on an x-axis across the piano keyboard start to resonate, adding another dimension to the spectrum of sound (see Figure 5).[58]

Because of the modern piano's tempered,[59] stretched[60] tuning, things do not exactly line up, however. Those z-axis Bs, F#s, D#s, As, and higher partials are not quite the same frequencies as their x-axis counterparts, and this is the key to the material acoustics Pärt is attuned to that I describe in terms of silence. The partials along the piano's z-axis strings are nearly harmonic because they produce near-integer multiples of a fundamental frequency in their vibrations. But those z-axis partials do not cleanly correlate with the tempered tuning of the x-axis piano strings. The nearly harmonic vibrational modes of z-axis strings differ slightly (and in an all-important way for my point here) from the pedal-down sympathetic vibration of x-axis strings tuned to a culturally partic-ular system of stretched tuning and equal temperament.[61]

You might imagine these two axes as different clocks, the z-axis clock running perpetually according to the oscillation of a pendulum and the x-axis clock being continually adjusted for daylight savings time, leap years, dead batteries, and personal whim. Quite quickly, the two clocks are in and out of sync, interfering, reinforcing, and grooving with each other, fixed and mov-able feasts. The same is true of the B strings' vibrating partials (z-axis) and the tempered sympathetic resonances (x-axis) of Pärt's pedal-down, spacious piano.

Figure 5. The Z and X axes on the piano.

An example of this is the seventh partial of B0, which is A3, vibrating at 216 Hz along the z-axis but tuned to 220 Hz along the x-axis in a tempered system (see Figure 5). Heard simultaneously, these interfering A3s produce a beat pattern of 4 Hz — a small fraction of the shimmering, pulsating spectrum of sound at the beginning of *Für Alina*. As the partials of the z-axis strings and the tempered resonances of the x-axis strings intersect, one can hear the vibrational difference that is the psychoacoustic ground of tintinnabuli's silence.

Looking at the spectrogram in Figure 6 of the opening Bs in *Für Alina*, one can see tintinnabuli's silence as formants — frequencies intensified by the overlapping vibrations of fundamentals, partials, and sympathetic resonances.

Figure 6. Spectrogram of the low Bs at the opening of *Für Alina.*

The breadth of the formants represents the frequency variations between the z- and x-axes. They are messy because there are more timbrally interesting, close-but-not-quite frequencies in play (play Audio Example 1, available at https://youtu.be/2kgHhkcNTzg).

The principal formants emerge here through the relationship of partials and resonances along the z- and x-axes. What is interesting and important to note is that these formants are not only derived from the notated Bs that open *Für Alina* but from the resonances of resonances within the multidimensional acoustic space of Pärt's piano. In Figure 6, for example, F#6 (around 1480 Hz) is the third partial of B4 (around 494 Hz), which vibrates as the fourth partial of the notated B2 (around 123 Hz). At the same time, F#6 is the second partial of F#5 (around 740 Hz), which vibrates as the third partial of B3 (around 247 Hz), which vibrates as the second partial of the notated B2. And so on and so forth through the other possible ways of arriving acoustically at this complex F#6 formant. My point is that this F#6 formant emerges through multiple vibrational and resonant pathways, each of which differs slightly in terms of vibrational frequency, producing the acoustically complex footprint of this single formant, a tiny detail of its tintinnabulation.

There is something not so tiny about this detail, though; something that extends to the phenomena of tintinnabulation more generally. The formants I am discussing are psychoacoustic phenomena emerging through the relationship of the z- and x-axes. There is a third axis, then, a third dimension that emerges from the vibrations along the z- and x-axes — a sounding y-axis produced through the intersections of z-axis partials and x-axis resonances (see Figure 7). This sonic dimension is an artifact of a listening subject's psychoacoustic perception of formants — an experience of the touches of string vibrations on their body as the material basis of sound.[62] In the reduced listening of tintinnabuli's silence, the y-axis is experienced through listening "in millimetres" to the intermaterial vibrations of sound.

To widen the aperture on this y-axis, I return briefly to the beginning of *Für Alina*, just as the low Bs are decaying and things are about to get more pungent when I play B4 and D6 in the next measure, which produces a D (notated pitch)/D# (y-axis) cross-relation (see Figure 2). Looking at the F#7 formant (around 2960 Hz) in the first phrase (see Figure 8), one can reverse engineer how the y-axis phenomenon materializes. Working off the pitch B4 (around 494 Hz) in the left hand, F#7 vibrates as its sixth partial (z-axis). Pedal down, that partial sets the F#7 string vibrating sympathetically (x-axis). Where things get interesting is in the perception of both these F#7s as a y-axis formant, since they vibrate at slightly different frequencies because of the piano's temperament, stretched tuning, and inharmonicity to produce a multidimensional, shimmering, subtly beating sound in constant transformation.

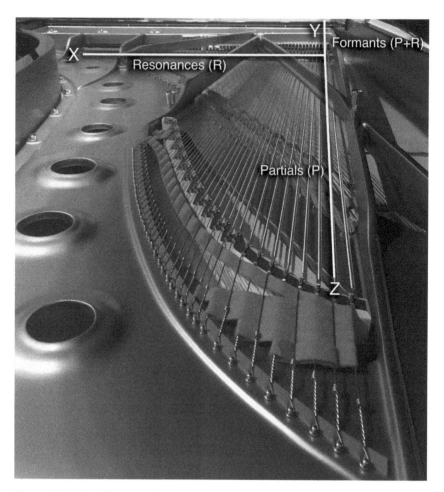

Figure 7. The Z, X, and Y axes on the piano.

To get at the acoustics of this F#7 formant fully, one would have to account for all the partials and resonant vibrations involved, which produce multiple F#7s along multiple y-axes, all of which constitute the formant. So the second partial of the left-hand B4, which is B5 (around 988 Hz) on the z-axis, makes the B5 x-axis string resonate, which produces a new F#7 (the third partial of B5) on a new z-axis. And the third partial of B4, which is F#6 (around 1480 Hz) on the z-axis, makes the F#6 x-axis string resonate, which produces a new F#7 harmonic on a new z-axis. All these versions of F#7, each with a slightly different frequency, intersect in the glittery, oscillating y-axis F#7 formant (see Figure 8).

I have disaggregated components of Pärt's material acoustics here to show

Figure 8. Spectrogram of *Für Alina*, mm. 1–2.

just a tiny sliver of tintinnabuli's sonorousness and silence and to make a simple point about the sonic turn in thinking about Pärt. The y-axis I write about is where tintinnabulation materializes between the sonic bodies of listeners, singers, and instruments: a possible intersection between vibrating material and intent listening to psychoacoustic phenomena. These essential acoustics exceed the medium of an instrument and the representation of sound in a musical score, and they reorient thinking about music around an attitude of listening, sonic embodiment, and the "timbral labor" of performers.[63] None of this, of course, is unique to Pärt or limited to the piano. Reduced listening is an attitude toward any kind of sound, and the z-, x-, and y-axis interactions within the piano extend to the interactions between singers and instrumentalists through intonation, timbre, vibrato, and vowel shape; between performers, listeners, and the resonant spaces in which music happens; and in the sensibilities of recording engineers, the fidelity of audio formats, and the qualities of playback technologies. However, a sonic approach to tintinnabuli that starts from vibration rather than notation illuminates and amplifies the impression of Pärt beholding silence in his musical designs. By way of reduction, quietude, textural transparency, slowness, and acoustic relationships set in relief, tintinnabuli gets out of its own way, so to speak; it clears space within a sonic spectrum to perceive the kind of silence I write about here. A complementary approach to algorithm- and text-driven treatments of tintinnabuli, then, is to turn the tables and consider tintinnabuli's bell-like acoustic phenomena as the grounds of Pärt's compositional choices, not the byproducts. Perhaps tintinnabuli's algorithms and forms are organized around a y-axis, meant to offer a sui generis experience of material vibration, meant to silence.

Epilogue: Tintinnabuli's Humanness

"Sound is a little piece of the vibrating world," writes Jonathan Sterne.[64] Just a "little piece" because sound is the vibration that is transduced and perceived by human beings. "We can say either that sound is a class of vibration that *might* be heard," Sterne continues, "or that it is a class of vibration that *is* heard, but, in either case, the hearing of the sound is what makes it."[65] Sound and silence are anthropocentric and relative to an individual's perceptual abilities (but not limited to the ear). The humanness of sound and silence, however one understands the human anthropologically, theologically, or biologically, make embodiment and sensation key concerns in thinking about tintinnabuli's wider resonances. In Pärt studies, tintinnabuli's material acoustics and listening practices, always coupled to listeners' culturally specific audile techniques[66] and social experience, are an essential part of conversations on Pärt's persona and biography, twentieth- and twenty-first-century music historiography, Christian theology, contemporary religion and spirituality, Estonian cultural history, and film and media.

In particular, I approach tintinnabuli's silence as a material basis for affective, somatic experiences of Pärt's music that cut across ideological categories like religious/spiritual/secular or Western concert music/popular music/film music to get at Pärt's broader publics.[67] Silence as I write about it is a counterpoint to the word in Pärt's music — the word as sound that communicates something beyond itself as part of language. As is well documented, the word is central to tintinnabuli's "rules of the game" and the messages Pärt's music mediates through his expansive engagement with texts from the Christian tradition.[68] In Bouteneff's meditation on Pärt's music, the possibilities of hearing the word in richly theological and spiritual or nonreligious ways exist side by side.[69] My claim here has been that there is a mode of listening, a way of being silenced by tintinnabulation that might suspend the question of religious/spiritual/nonreligious experience in the encounter with Pärt's material acoustics per se. When the word is foreign, not understood, or not heard, tintinnabuli's silence grounds listeners in the humanness of sound. When Pärt's instrumental music offers no immediate words, its material vibrations and psychoacoustics ground listeners in fleeting moments of attunement to their embodied perception. And when listening is reduced to tintinnabuli's acousmatic silence, the matter of "meaning" in Pärt's music is the sound event itself. Silent, and vibrating, tintinnabulation is profoundly human; it musicalizes acoustic phenomena through their materialization in human audition and perception.[70]

Notes

1. The epigraphs to this chapter are from Eric Marinitsch, "As Purely, Cleanly, and Simply as Possible: Interview with David James," https://www.arvopart.ee/en/arvo-part/article/as-purely-cleanly-and-simply-as-possible; and Geoffrey J. Smith, "Sources of Invention: An Interview with Arvo Pärt," *Musical Times* 140, no. 1868 (1999): 21, https://www.arvopart.ee/en/arvo-part/article/sources-of-invention/.

2. Partials are modes of string vibration in the piano that, for an ideal string, produce integer ratios of a fundamental frequency (f_0) in a harmonic series ($f_1=2f_0$, $f_2=3f_0$, and so on), which is the basis of so-called Pythagorean tuning (but not the tempered and stretched tuning of the modern piano). Because of the mass and stiffness of piano strings, however, partials diverge slightly from ideal harmonic behavior, producing frequencies a few Hz higher in the upper partials — a phenomenon that enhances the points I make here. For more on the inharmonic nature of the piano, see Nicholas J. Giordano, *Physics of the Piano* (New York: Oxford University Press, 2010), 57–62.

3. Cents are logarithmic units for measuring intervals developed by Alexander J.

Ellis. In Ellis's system, an octave is divided into 1,200 cents, where a semitone is measured by 100 cents. Alexander J. Ellis, "On the Musical Scales of Various Nations," *Journal of the Society of Arts* 33 (1885): 485–527.

4. Paul C. Jasen, *Low End Theory: Bass, Bodies, and the Materiality of Sonic Experience* (New York: Bloomsbury, 2016), 66–71.

5. For Helmholtz, who advised Theodore Steinway on piano construction, the heuristic value of the pedal-down piano in theorizing complex sounds and vowel formants as the combination of partials was strong: "Raise the dampers of a pianoforte so that all the strings can vibrate freely, then sing the vowel *a* in *father*, *art*, loudly to any note of the piano, directing the voice to the sounding board; the sympathetic resonance of the strings distinctly re-echoes the same *a*. On singing *oe* in *toe*, the same *oe* is re-echoed. On singing *a* in *fare*, this *a* is re-echoed. For *ee* in *see* the echo is not quite so good. The experiment does not succeed so well if the damper is removed only from the note on which the vowels are sung. The vowel character of the echo arises from the re-echoing of those upper partial tones which characterize the vowels. These, however, will echo better and more clearly when their corresponding higher strings are free and can vibrate sympathetically. In this case, then, in the last resort, the musical effect of the resonance is compounded of the tones of several strings, and several separate partial tones combine to produce a musical tone of a peculiar quality." Hermann von Helmholtz, *On the Sensations of Tone as a Physiological Basis for the Theory of Music*, 2nd English ed., trans. Alexander J. Ellis (1885; New York: Dover, 1954), 61.

6. Acoustic beats are the interference patterns of slightly different frequencies sounding simultaneously. Beats are perceived as periodic changes in loudness (amplitude) according to the rate at which frequencies vary. Simultaneously sounding frequencies of 222 Hz and 221 Hz, for instance, produce beats at a frequency of 1 Hz. Beat frequencies between 10 Hz and "as low as a few tenths of a Hz are perceptible" by the human auditory system. Giordano, *Physics of the Piano*, 21.

7. For more on the relationship of partials, beating, pedal, and decay, see Heidi-Maria Lehtonen et al., "Analysis and Modeling of Piano Sustain-Pedal Effects," *Journal of the Acoustical Society of America* 122, no. 3 (2007): 1787–97.

8. See Felipe Pinto d'Aguiar, "Tintinnabulation under the Microscope," https://www.academia.edu/5814272/Arvo_P%C3%A4rt_Musical_Analysis, for a psychoacoustic approach to Pärt's sound focusing on combination tones and the "hidden voices" in *Für Alina* and other Pärt pieces. D'Aguiar's work resonates with what I write about here in a commitment to approach "music from sound and the listening process rather than from the score." I do not take up the psychoacoustic phenomenon of combination tones here, which forms the basis of d'Aguiar's work.

9. The Goeyvaerts Trio speak about and demonstrate their work with just intonation and vibratoless playing here: https://www.youtube.com/watch?v=ctuts SGWsrM.

10. Philippe Grisar, liner notes to the Goeyvaerts Trio's recording of Pärt's *Stabat Mater* (Challenge Classics CC72616).

11. Nina Sun Eidsheim, *Sensing Sound: Singing and Listening as Vibrational Practice* (Durham, NC: Duke University Press, 2015), 163.

12. Michel Chion, *Sound: An Acoulogical Treatise*, trans. James A. Steintrager (Durham, NC: Duke University Press, 2016), 16.

13. For Pierre Schaeffer's elaboration on modes of listening, see his *Treatise on Musical Objects: An Essay across Disciplines*, trans. Christine North and John Dack (Berkeley: University of California Press, 2017), 103–16.

14. Schaeffer, *Treatise on Musical Objects*, 115.

15. Schaeffer, *Treatise on Musical Objects*, 213. Schaeffer is quick to point out that, as a mode of attention, reduced listening is difficult and, in his words, "antinatural": "Most of the time, as we have seen, my listening targets *something else*, and I hear only indicators or signs. Even if I attend to the sound object, my listening will for an initial period be *referential listening*. In other words, there will be no point in being interested in the sound itself; at first I will still be incapable of saying anything else about the sound than 'it's a galloping horse,' 'it's a creaking door,' 'it's a G-flat on the clarinet,' 'it's 920 periods per second,' it's 'Hello, hello.' The cleverer I have become at interpreting these sound indicators, the more difficult it will be for me to hear the objects. The better I understand a language, the more difficult it will be to *perceive it with my ear*" (212).

16. Chion, *Sound*, 170. For Chion's take on Schaeffer's modes of listening, see 169–211.

17. Perceptually, but not ontologically, in my view, since the "audit," what John Mowitt calls an "analogue to *gaze* in the auditory or sonic domain," is the psychoacoustic event of this silence. John Mowitt, *Sounds: The Ambient Humanities* (Berkeley: University of California Press, 2015), 4. On the ongoing ontology/perception debates in sound studies, see Christoph Cox, "Sonic Realism and Auditory Culture: A Reply to Marie Thompson and Annie Goh," *Parallax* 24, no. 2 (2018): 234–42; Annie Goh, "Sounding Situated Knowledges: Echo in Archeoacoustics," *Parallax* 23, no. 3 (2017): 283–304; Brian Kane, "Sound Studies without Auditory Culture: A Critique of the Ontological Turn," *Sound Studies* 1, no. 1 (2015): 2–21; and Marie Thompson, "Whiteness and the Ontological Turn in Sound Studies," *Parallax* 23, no. 3 (2017): 266–82.

18. Brian Kane, *Sound Unseen: Acousmatic Sound in Theory and Practice* (New York: Oxford University Press, 2014).

19. Schaeffer, *Treatise on Musical Objects*, 69.

20. Chion, *Sound*, 201.

21. See Kane, *Sound Unseen*, 5.

22. Cf. Julian Henriques, *Sonic Bodies: Reggae Sound Systems, Performance Techniques, and Ways of Knowing* (New York: Continuum, 2011); and Jim Sykes, "Ontologies of Acoustic Endurance: Rethinking Wartime Sound and Listening," *Sound Studies* (2018), doi:10.1080/20551940.2018.1461049.

23. See Philip V. Bohlman, "Music as Representation," *Journal of Musicological Research* 24, nos. 3–4 (2006): 205–26.

24. See Ana María Ochoa Gautier, "Silence" in *Keywords in Sound*, ed. David Novak and Matt Sakakeeny (Durham, NC: Duke University Press, 2015), 183–92.

25. David Novak, "Playing Off Site: The Untranslation of *Onkyô*," *Asian Music* 41, no. 1 (2010): 36–59.

26. John Biguenet, *Silence* (New York: Bloomsbury Academic, 2015).

27. One could add two other aspects to this silence: (1) the effects and affects of vibration over durations shorter than the temporal resolution of the human auditory system (millisecond durations whose envelope is shorter than the time it takes for the auditory system to transduce vibration into electrical energy; the short time vibration is felt before it is heard), and (2) the haptic effects of infrasound or ultrasound. See Steve Goodman, *Sonic Warfare: Sound, Affect, and the Ecology of Fear* (Cambridge, MA: MIT Press, 2010); Jasen, *Low End Theory*; and Schaeffer, *Treatise on Musical Objects*, 150–60.

28. See Leopold Brauneiss, *Arvo Pärdi tintinnabuli-stiil: arhetüübid ja geomeetria*, trans. Saale Kareda (Laulasmaa: Arvo Pärt Centre, 2017); Paul Hillier, *Arvo Pärt* (New York: Oxford University Press, 1997), 86–97; and Andrew Shenton, *Arvo Pärt's Resonant Texts: Choral and Organ Music, 1956–2015* (Cambridge: Cambridge University Press, 2018), 33–46.

29. "Vaikus on tarkus. Vaikus — sa pead suhtlema väga suure aukartusega. Vaikus mõistab kohut meie peale, mitte ainult iga sõna peale ja heli peale, vaid ka iga mõtte peale, mis võiks olla sisuliselt patt, ebapuhas." Quotation taken from Immo Mihkelson's radio documentary *Arvo Pärt 70*, part 14 ("Mõtteid Arvo Pärdi loomingust"), section 4 ("Vaikus"), 2005.

30. Wolfgang Sander, liner notes to *Tabula rasa* (ECM New Series 1275).

31. Immo Mihkelson, *Arvo Pärt 70*, part 7 ("Vaikus, mis muutis maailma. Uus helikeel. Tõus ja lahkumine"), section 3 ("Tabula Rasa"), 2005.

32. Ines Rannap, "Muusikasündmus TPI aulas," *Sirp ja Vassar*, October 21, 1977; and Aurora Semper, "Rahvusvahelise muusikapäeva tähistamiseks," *Rahva Hääl*, October 27, 1977. Cited in Kevin C. Karnes, *Arvo Pärt's Tabula Rasa* (New York: Oxford University Press, 2017), 126.

33. This issue of *Teater. Muusika. Kino.* also featured an overview of critical praise from the *Neue Zeitschrift für Musik*, *Gramophone*, *Frankfurter Allgemeine Zeitung*, the *Times* (London), the *New York Times*, *Helsingin Sanomat*, and others for Pärt's second ECM album *Arbos* (1987) and a reproduction of the Continuum ensemble's pioneering 1984 Pärt retrospective concert in New York City.

34. Leo Normet, "The Beginning Is Silence," *Teater. Muusika. Kino.* 7 (1988): 19–30; Ivalo Randalu, "Arvo Pärt novembris 1978," *Teater. Muusika. Kino.* 7 (1988): 48–55; and Merike Vaitmaa, "Tintinnabuli: eluhoiak, stiil ja tehnika," *Teater. Muusika. Kino.* 7 (1988): 37–47.

35. "All music emerges from silence, to which sooner or later it must return." Hillier, *Arvo Pärt*, 1.

36. See, for instance, Pärt's interview with Tom Huizenga on NPR in "The Si-

lence and Awe of Arvo Pärt," June 2, 2014, https://www.npr.org/sections/deceptive cadence/2014/06/02/316322238/the-silence-and-awe-of-arvo-p-rt.

37. Laura Dolp, "Ethos and the Industry of Culture," in *Arvo Pärt's White Light: Media, Culture, Politics*, ed. Laura Dolp (Cambridge: Cambridge University Press, 2017), 97–121.

38. Karnes, *Arvo Pärt's Tabula Rasa*.

39. Shenton, *Arvo Pärt's Resonant Texts*. Also see Shenton's chapter in this volume.

40. See Toomas Siitan's lecture "Music and Silence" ("Muusika ja vaikus"), one of nine public lectures delivered as part of Pärt's term as professor of fine arts at the University of Tartu in 2013–2014: http://www.uttv.ee/naita?id=19801.

41. Peter C. Bouteneff, *Arvo Pärt: Out of Silence* (Yonkers, NY: SVS, 2015), 105–6.

42. Cf. Judith Becker, *Deep Listeners: Music, Emotion, and Trancing* (Bloomington: Indiana University Press, 2004).

43. Henriques, *Sonic Bodies*.

44. Jasen, *Low End Theory*; and Robert Fink, "Below 100 Hz: Toward a Musicology of Bass Culture," in *The Relentless Pursuit of Tone: Timbre in Popular Music*, ed. Robert Fink et al. (New York: Oxford University Press, 2018), 88–116.

45. See Marguerite Bostonia, "Bells as Inspiration for Tintinnabulation," in *The Cambridge Companion to Arvo Pärt*, ed. Andrew Shenton (Cambridge: Cambridge University Press, 2012), 128–39; and Hillier, *Arvo Pärt*, 18–23, 86–87.

46. Cf. Karnes, *Arvo Pärt's Tabula Rasa*, 37–44.

47. Shenton, *Arvo Pärt's Resonant Texts*, 35.

48. Paul Hillier writes something similar: "The sonority which accumulates is intrinsically clear yet contains overtones and undertones far more dense than the notes on paper would suggest" (*Arvo Pärt*, 86).

49. Christopher Small, *Musicking: The Meanings of Performing and Listening* (Middletown, CT: Wesleyan University Press, 1998).

50. "Hõlmab . . . kahe muusikalise liini üheaegselt kõlamise vahekorda — nende helide suhteid." Quotation taken from the *"Tintinnabuli* — Style, Method, or Mindset?" *("Tintinnabuli* — stiil, metood või mõtteviis?") episode of Immo Mihkelson's *Arvo Pärt 70* radio documentary.

51. Formants are focused amplitude peaks in the frequency spectrum of a sound. In linguistics, the distribution and strength of formants determine the identities of vowels; for instruments and voices, formants determine timbre — why a pitch with the same fundamental frequency sounds different on a piano and a guitar, for example.

52. See Thomas D. Rossing, *Science of Percussion Instruments* (River Edge, NJ: World Scientific, 2000), 67–69. Rossing comments: "One of the interesting characteristics of chimes [tubular bells] is that there is no mode of vibration with a frequency at, or even near, the pitch of the strike note one hears" (68). For a compar-

ison of the acoustics of tuned and untuned bells, see Bostonia, "Bells as Inspiration for *Tintinnabulation*," 136–37.

53. Hillier, *Arvo Pärt*, 20.

54. "Kus niimoodi kontsentreeruda iga heli peale, nii nagu igal rohuliblel oleks nagu lille staatus . . . pisikeses helis, pisikeses fraasis näha midagi rohkemat kui ainult see must klahv ja see valge klahv." Quotation from Dorian Supin, *24 Preludes for a Fugue (Arvo Pärt. 24 prelüüdi ühele fuugale)* (F-Seitse, 2002).

55. "Mis maksab üks heli või sõna? Need tuhanded, mis meie kõrvust mööda voolavad, on teinud tuimaks meie vastuvõtuaparaadi. Hoolikalt suhtuda igasse helisse, sõnasse, teosse." https://www.arvopart.ee/esimese-tintinnabuli-kontserdi -kavalehest/.

56. Geoffrey J. Smith, "Sources of Invention: An Interview with Arvo Pärt," *Musical Times* 140, no. 1868 (1999): 22.

57. The lowest partials of vibrating B strings materialize in nearly harmonic proportions. B is the second and fourth partial with ideal frequency ratios of 2:1 and 4:1, respectively; F# is the third partial with an ideal frequency ratio of 3:1; and D# is the fifth partial with an ideal frequency ratio of 5:1.

58. Helmholtz explains a popular demonstration of sympathetic resonance along what I am calling the x-axis on the piano: "Gently touch one of the keys of a pianoforte without striking the string, so as to raise the damper only, and then sing a note of the corresponding pitch forcibly directing the voice against the strings of the instrument. On ceasing to sing, the note will be echoed back from the piano. It is easy to discover that this echo is caused by the string which is in unison with the note, for directly the hand is removed from the key, and the damper is allowed to fall, the echo ceases." Helmholtz, *On the Sensations of Tone*, 38–39.

59. Equal temperament is a logarithmic tuning system designed to space all twelve semitones equally across the octave, thereby addressing the issue of the Pythagorean comma and enharmonic inequivalence in a tuning system based on perfect fifths with a ratio of 3:2. The issue with Pythagorean tuning is that twelve perfect fifths, which would generate all the semitones in an octave, are not equivalent to seven octaves; they do not produce an enharmonic system in which C and B# are the same frequency. Twelve perfect fifths $(3:2) = 1.5^{12} = 129.746 . . . $; seven octaves $(2:1) = 2^7 = 128$. See Giordano, *Physics of the Piano*, 28–33.

60. Octave stretching (tuning the octave to a frequency ratio slightly greater than 2:1, known as the Railsback stretch) compensates for the fact that "string stiffness shifts the partials of a vibrating piano string to frequencies higher than those of a perfect harmonic spectrum," producing an undesirable level of acoustic interference. Giordano, *Physics of the Piano*, 60.

61. See Giordano: When tuning with imperfect, tempered fifths, "the third harmonic of C4 is 0.44 Hz higher than the second harmonic of G4 [both of which should be G5]. These two harmonics will thus give beats . . . with a beat frequency of 0.44 Hz, which is quite noticeable." Giordano, *Physics of the Piano*, 31.

62. Helmholtz's poetic thinking (modified by later researchers) about resonance

through the "nervous piano" bears directly on this perception. "Now suppose we were able to connect every string of a piano with a nervous fibre," Helmholtz writes, "in such a manner that this fibre would be excited and experience a sensation every time the string vibrated. Then every musical tone which impinged on the instrument would excite, as we know to be really the case in the ear, a series of sensations exactly corresponding to the pendular vibrations into which the original motion of the air had to be resolved. By this means, then, the existence of each partial tone would be exactly so perceived, as it really is perceived by the ear. The sensations of simple tones of different pitch would under the supposed conditions fall to the lot of different nervous fibres, and hence be produced quite separately, and independently of each other." Helmholtz, *On the Sensations of Tone*, 129. For a critique of Helmholtz's incomplete theory of resonance as it relates to his musical aesthetics and broader modern epistemologies, see Veit Erlmann, *Reason and Resonance: A History of Modern Aurality* (New York: Zone, 2010), 232–70.

63. Eidsheim, *Sensing Sound*, 135.

64. Jonathan Sterne, *The Audible Past: Cultural Origins of Sound Reproduction* (Durham, NC: Duke University Press, 2003), 11.

65. Sterne, *The Audible Past*, 11.

66. See Sterne, *The Audible Past*; and Ana María Ochoa Gautier, *Aurality: Listening and Knowledge in Nineteenth-Century Colombia* (Durham, NC: Duke University Press, 2014).

67. I touch on these themes in two earlier pieces of writing: "Arvo Pärt and the Idea of a Christian Europe: The Musical Effects and Affects of Post-Ideological Religion," in *Resounding Transcendence: Transitions in Music, Religion, and Ritual*, ed. Jeffers Engelhardt and Philip V. Bohlman (New York: Oxford University Press, 2016), 214–32; and "Perspectives on Arvo Pärt after 1980," in *The Cambridge Companion to Arvo Pärt*, ed. Andrew Shenton (Cambridge: Cambridge University Press, 2012), 29–48.

68. See Hedi Rosma et al., eds. *In Principio: The Word in Arvo Pärt's Music* (Laulasmaa: Arvo Pärt Centre, 2014).

69. Bouteneff, *Arvo Pärt*, 25–47.

70. Special thanks to Peter Bouteneff, Kevin Karnes, and Rob Saler for their comments and improvements on this chapter.

9

Medieval Pärt

Andrew Albin

The title of this chapter calls up a familiar image of Arvo Pärt and his music: Pärt as mystic composer out of step with our time, strange stowaway from a remote, incense-filled era of history, his music harboring against all odds in the contemporary musical-cultural landscape. This image of the medieval Pärt depends partially on stylistic qualities of his music that hark back to soundworlds we recognize as medieval,[1] partially on our popular narratives about the medieval past, so often represented as a Dark Ages less interested in scientific and cultural advancement than in the ecstatic haze of religious devotion. Though false, this influential version of the Middle Ages stands both as a beacon of difference and a locus of desire: It is the awe-filled, childlike historical Other against which our rational, post-Enlightenment selves are defined, whose unsullied wonder at the world becomes an object of longing in the face of the failed promises of modernity.[2]

It is to this version of the Middle Ages that listeners and scholars often recur in an effort to explain the appeal of Pärt's music, which departs so distinctively from what our ears have been taught to hear as the sounds of modernity and musical progress. Critics have likened the harmonic stability and structural architecture of Pärt's music to the echoing spaces of the medieval cathedral.[3] Pärt reportedly pored over medieval and renaissance repertoires — plainchant, Machaut, Josquin — in his search for a new compositional style during his eight years of silence. The name by which this new style came to be known, "tintinnabuli," is redolent of medieval church bells tolling out the hours of the day before the advent of mechanized clock time. Medievalizing tropes equally shape Pärt's popular image: Wolfgang Sandner likens Pärt's compositions to "the hesychastic prayers of a musical anchorite,"[4] mixing Eastern and Western religious terminology to compare Pärt's output to that of a solitary hermit sunk in silent meditation. Nora Pärt makes the medieval connection explicit, remarking on how Western media eagerly seized upon her husband as "an exotic being: mystic, monk, beard, medieval vocabulary, detached from the world, etc."[5] The stories we tell ourselves about the Middle Ages, the language we use to describe Pärt's music, and the public image built around Pärt, supported by ECM's marketing, dovetail neatly to evoke this distant historical past in the immediate sonorous present of listening to Pärt's compositions.

What underlies this evocation and the aesthetic, cultural, and historical attractions it arouses? To begin answering this question, we do well to notice those attitudes toward the process of history that support this image of the medieval Pärt. To return to the formulations with which I began: If Pärt's music is "out of joint" with our modern day, then our modern day is understood to be an integral, organic body, a whole that his music dislocates. If Pärt's music is a stowaway from the medieval past, the ship of history that it hid upon voyages ineluctably toward the secure port of our present. We become the inevitable endpoint of chronology's trajectory. Time flows in one direction, pressing ever onward; all that has come before us points toward us and anticipates us, preparing the way for our arrival. From this point of view, the narrative of history is a narrative of progress, familiar from our history textbooks, music and otherwise, and we are the lucky beneficiaries who stand at its leading edge.

If this is how history works, if history shuttles steadily toward our triumphant now, then there can be little place in the present for an artistic voice like Pärt's that resists the values of the avant garde and fits poorly into our received narratives of stylistic evolution. It must come as little surprise, then, that Pärt's music has been disparaged as conservative or recidivist:[6] It does not fit comfortably with our historical teleology, so it must be located elsewhere, relegated to a past we have left behind. The medieval thus comes to mark Pärt's disruption of the logic of progressive modernity and teleological historicity; it becomes a rhetorical means of framing, and thereby containing, his music's confident, unnerving refusal to heed the modish pull of musical pioneership.

This is not the only way to understand how history works, however. Scholars of the Middle Ages in particular recognize that time need not flow only forward: We study how medieval people observed and lived multiple, overlapping temporalities other than that of the ticking clock.[7] Seasonal, laboring, liturgical, typological, and eschatological time crisscrossed the onward march of a lifespan, sweeping human beings into larger, multidirectional organizations of time as time spooled out in either direction from the zero point of Christ's birth. Moreover, medievalists have also recognized how the Middle Ages have a funny way of returning, of breaking through the surface of subsequent historical periods, often at moments of crisis or transformation, to shape how these later moments in history understand their identity, their lineage, and their future. Heeding these irruptions serves as a helpful reminder to call our self-assured presentism into question and decenter our hubristic conviction that we stand at the leading edge of historical progress's unquestioned good. In doing so, the historical past is no longer relegated to the netherworld of temporal remoteness; rather, it remains energetically in circulation, both as narrative trope that later historical moments can use to various cultural and political ends and as structural force that lingers, recedes, accompanies, and breaks into later historical moments in variably overt and subtle ways.[8] I see Arvo Pärt, his music, and the discourse surrounding him and his music as one such latter-day irruption of the medieval into a subsequent historical moment.

It is with this more polymorphous sense of time in mind that the present chapter proceeds.[9] Contrary to what this chapter's title might be taken to imply, I do not wish to enumerate the qualities that mark Pärt or his music as medieval stowaways — we need look no further than the man's birth certificate to determine that he is in fact modern like the rest of us. Rather, I want to bracket the assumptions of teleological chronology and consider how the medieval reinstates itself in our present moment through Pärt, how Pärt's music creates sounding environments in which modes of experience familiar to the Middle Ages might resurface and expand. To be clear, this is not to suggest that Pärt's music opens some sonorous portal by which we may experience music or the self or the world or the divine in the same way that medieval people experienced those things six or more centuries ago. Medieval experiences can never be ours, because the experiences of medieval people were as diverse as ours and were shaped by cultural contexts fundamentally different from ours — indeed, the fantasy of reexperiencing the Middle Ages through Pärt's music is arguably a product of our unidirectional narratives of history. Rather, I want to reflect on how medieval perspectives on the materials of Pärt's compositions — words, texts, bodies, sounds, silences — might help us understand our modern-day sensorial and spiritual encounters with his music better. To put it more simply, I want to ask not how Pärt is medieval but what the medieval helps us hear in Pärt.

Musical Mysticism: Richard Rolle and Medieval Textual Culture

To do the work of listening to Pärt's music through medieval ears,[10] we might profitably begin by seeking out historical analogues whose own creative output echoes, in orientation and effect if not in musical particulars, that of our Estonian composer. Numerous medieval candidates come to mind, but I wish to focus on one fourteenth-century figure in particular: Richard Rolle of Ham-pole, a hermit given to outwardly silent and solitary prayer, a mystic deeply invested in the sounds of spiritual music, and the most widely read author of England's late Middle Ages. Rolle was the premiere sound artist of his day: His texts in Latin and Middle English are full of evocative song lyrics, unparalleled sound effects, effusive descriptions of musical audition, and sustained meditations on the spiritual dimensions of sound and silence. Music founds his highest contemplative flights; for him, union with God amounts to conversion into harmonious musical consonance that affects the entire person, transforming the experience of hearing into much more than the perception of sound waves propagated in air. Though scholarly opinion has customarily disparaged him as a second-rate mystic because of his orientation toward the senses, Rolle held far greater appeal for English spiritual life than any other canonical English mystic, such that each of his successors reacts in one way or another to his sensory spirituality and its wide cultural impact. In many ways, then, Rolle presents a particularly apposite figure to put into conversation with Arvo Pärt: Despite divides in Christian confession, artistic medium, and historical epoch, the two men are strikingly kindred spirits, sharing devotional proclivities, creative inclinations, and a public reception equal parts ambivalent and enthusiastic.

The core of Rolle's popular appeal lay in the emotional and sensorial intensity of his writings. Soon after ending his studies at Oxford and rebelliously adopting the eremitic lifestyle,[11] Rolle experienced a trio of spiritual sensations that would become the hallmark of his mysticism: *fervor*, a heartfelt heat warming the body from within; *dulcor*, a sweetness filling the mouth and nose; and *canor*, a state of constant musical coparticipation with the heavenly choirs and the apex of Rolle's spiritual ascent. These experiences formed the nucleus of the hermit's writings, primarily in Latin; his literary career can be seen as a lifelong effort to explain these sensations, convince ecclesiastic skeptics of their verity, and encourage others to likewise seek them. Rolle eventually settled in the town of Hampole, the site of a poor Cistercian nunnery, where he served as spiritual advisor to local religious women for whom he wrote a number of works in Middle English. When he began to perform miracles from beyond the grave after his death on September 29, 1349, the hermit's fame quickly spread, demand for his texts grew, the fortunes of Hampole Priory took a distinct turn for the better, and a popular cult arose that lasted well into the Renaissance, though it never gained enough traction for his prospective sainthood to be ratified.[12]

Rolle's writings steadily grew in popularity during the later fourteenth and fifteenth centuries, a period when emotional, embodied styles of Christian

piety flourished and vernacular authorship and devotional readership dramatically expanded. As the hermit's writings reached new audiences — not just the
learned, religious, and predominantly male readers for whom they had origi-
nally been authored but now lay, secular, and female readers hungry for first-
hand sensory experience of the divine — the audition of angelic song appears
to have become a familiar, if still remarkable, experience. Thanks to Rolle's
writings, England's Christian faithful were hearing angelic music as reward for
their devotion; efforts were made both to affirm and contain this potentially
dangerous audition.[13] English testamentary bequests of Rolle's works during
the late Middle Ages are widespread, exceeded in number only by the en-
trustment of biblical books. We are left today with at least 470 medieval manuscripts containing Rolle's works, amounting to a nearly complete record of
his literary output, an exceedingly rare case among medieval English authors
and clear testament to the man's impact on the popular religious culture of
his day.[14]

I wish to propose, then, that Richard Rolle, his spirituality, and his artistry
may be able to teach us something about Arvo Pärt and his music, once we
pay attention to the forms of hearing, musical and otherwise, that Rolle and
his writings sponsor. We learn a great deal about Rolle's own experience of
spiritual song from the autobiographical account of his inaugural audition of
canor in one of his most popular Latin treatises, the *Incendium amoris*, or *Fire
of Love*.[15] "While I was sitting in that same chapel," he writes,

> and singing the psalms the best I could at night before supper, I heard
> above me the ringing of psalm-chanters, or rather, of singers. And
> while I was occupied with heavenly matters, praying with all my de-
> sire, I cannot explain how, I next felt in me a choiring of song, and
> received a most delectable, heavenly harmony that dwelt with me
> in my mind. For my thought was transformed continually into cano-
> rous song, and I had odes as it were by meditation, and in those same
> prayers and in psalm-chanting I uttered the same sound. Thenceforth,
> continuously singing what before I had spoken, I burst out overflowing
> from inward sweetness, but privately, and only before my Maker.[16]

Rolle first hears *canor* in a liturgical context: Chanting psalms in a twilit
chapel, he initially takes the heavenly sound he hears to be that of fellow
psalm-chanters, only to realize the mistake as his being erupts with a new kind
of harmony, an internal music distinct from the song produced by his outward
voice, audible only to him and his God. Inward and outward song are contrasted at the same time that they resemble each other; the institutional community of the monastic choir cedes its functional role to a saintly community

of angels that far exceeds it.[17] Indeed, after he receives *canor*, Rolle prefers to celebrate Mass not in church but in the silence of his cell, a strikingly private turn in a Western Christian tradition where community membership was in many ways axiomatic to the constitution of self.[18]

In his most virtuosic Latin treatise, the *Melos amoris*, or *Melody of Love*, Rolle describes *canor* in significantly greater detail: It is inexpressible in words; it delights the soul through the senses; it is heaven sent and felt directly by the soul, not mediated through external sense organs. Rolle urges his audience to unplug their inner ears and open their inner eyes, explaining that *canor* has the power to screen off the five corporeal senses from the world. The contemplative who has achieved union with the divine is thus "like a chorister enclosed in the canticle, vital within and without. Her senses sustain their inmost attentions and aim at their only object so efficiently and avidly to the absolute exclusion of every other thing that they're unable to ebb away or avert to anything else."[19] Divine song erects the walls of an inner architecture, blocking out the world not by denying sense perception but rather by flooding it with so much heavenly stimulus that no room is left for the world's carnal lures. The human form becomes a tightly sealed chamber open only to God; the mystical body becomes a kind of spiritual resonator giving back the sounds it receives, its living structures echoing songs of praise, not unlike the cathedralesque acoustic space in which Rolle first heard *canor*.

Of course, Rolle's readers can only take him at his word—there is no way to verify whether the hermit actually experienced the sensations he claims. All we have to go on are his texts, which the hermit crafts with this problem in mind. At the very end of the *Incendium amoris*, Rolle is unambiguous about what his texts can and cannot offer the aspiring contemplative: "Lovers of the world can indeed know the words or verses of our songs, but not the singing of our verses; for they read the words but they cannot supply the note or the tone or the sweetness of these odes."[20] Though texts may offer words that lead a reader toward the audition of angelic song, they cannot grant access to *canor* directly; that allowance is the sole prerogative of God. What, then, are we meant to do with the insistent, rhythmic sonority of a passage like the following, from chapter 17 of the *Melos amoris*?

Frustra fundantur falsi fideles quia funditus finietur fiducia fenerantis, et fumo inferni ficti ferientur et omnes utique umbra honoris operti ut appareant in aulis avaris. Fervebunt fetentes formidine futura; formosus et fortis in feno falluntur et ideo imbuti impio instinctu fervore felici numquam fruentur quia federati fuerunt in factis falsorum ut fixi in fervore finiendi favoris feruntur cum furibus facibus frementes: horum fornax fetidus fauces iam fringet, nam fugiunt fidem famamque fugant; sic filii feroces firmantur fortiter ut fundum furencium penetrent post pauca et penas percipiant perpetuo perdurantes.

[The false faithful lay a futile foundation: financiers' fidelity will fully fail and fumes from the inferno will flog all the fraudulent upstarts obscured under the umbrella of honors that attracted them all into avarice's abbeys. The fetid will fall feverish in future fright; formidable foppish fellows are fooled with chaff and, imbued as they are with impious instinct, fortunate fervor will forego them; they fell confederate with falsifiers' feats like flunkies fixed in a fury for foisting favors, and they're fetched off with riffraff to fret in firebrands' flames; the foul furnace now fractures their pharynx, for they flee their faith and flout their fame to flight; these ferocious offspring are more firmly fated therefore to promptly pierce the paroxysmal pit and perceive perpetually perduring punishments.][21]

This passage is not unusual for the lengthy *Melos*, which alliterates to varying degrees from start to finish, reaching strings of up to forty words in a row with the same initial consonant, often observing a flexible four-stress rhythm. While most critics have treated this verbal cacophony representationally, regarding it as Rolle's attempt to transcribe the music of *canor* on the page, the hermit's own clear pronouncements to the contrary indicate that this cannot be. As I argue elsewhere, I believe we are on firmer ground when we understand Rolle's alliterative style gesturally: If a text cannot give its reader angelic song, it can activate and valorize sonority as a powerful arena of elevated experience, distanced from the directed semiotics of language but still charged with significance. Rolle's alliterative choices are independent of, even aleatoric to the meanings of the words on the page; his insistently patterned phonemes hover outside his words' signification, just as the experience of *canor* hovers outside the grasp of the aspiring contemplative reader. In this way, Rolle's alliteration generates desire for sound that is *not* there through sound that *is* there. Words and sounds collaborate in the evocation of powerful states of longing that stretch out for musical transformation and admission to the spiritual choir that renders praise to God, in whose Trinity three and one perfectly harmonize.[22]

The sonorous desiring effects the *Melos amoris* so powerfully produces rely as much on the expectations, literacies, and ideologies readers brought to the experience of reading in late medieval England as it does on the specific formal structuration of sound and text, music and language, body and soul Rolle's treatise so compellingly constructs. Late medieval textual cultures still predated the print technology, levels of education, and spheres of privacy that allowed silent reading to become culturally normative. Research into reading practices of the period suggests that, given the relative scarcity and expense of books and varying kinds and degrees of literacy, it was common practice for books to be read aloud, often in a social setting, with the expectation of digression and discussion.[23] Paper came into increasing use in the fifteenth century, but parchment was still the most common material out of which books were made during Rolle's lifetime. For him, then, the folios of a manuscript, the technological medium for which he understood his texts to be destined, would have comprised so many hides of so many animals, once living creatures whose skin touches our skin when we turn the page, whose flesh carries voices we reanimate when we read their tattooed words aloud.[24]

Moreover, if someone wanted to possess Rolle's *Incendium amoris* or *Melos amoris*, he would need to locate a manuscript containing the text, pay a scribe to copy it out, or copy the text by hand himself, word by word. A scribe's hand thus leaves its distinctive signature on the surface of the parchment page, functioning as an overdetermined metonym for a speaking voice whose sound the manuscript's owner may well have known.[25] This effect multiplies when one scribe hands off his work to another mid-text, when readers ink their notes and commentary in the margins, and when books are broken apart so their pieces can be joined together with new companions — all common textual practices for the period. Add to this the use of incipits, conventional verbal tags, to signal a vast memorized repertoire of liturgical verses and melodies and the occasional mixing of text and music notation together in the same book, and the act of reading a medieval manuscript emerges as a thoroughly embodied, thoroughly sonorous performance through which multiple voices might sound and interact in a dense play of presences and absences.

Still, we may wish to ask: Are these voices actually sonorous? Do they make sound, in the moment of human encounter with the book? From a purely acoustic perspective, these voices embedded in the manuscript are indeed silent, figural, voices in notion only, their sound only audible in the recondite cell of our imagination. Yet once we leave physics to venture into psychoacoustics, phenomenology, spirituality — realms of experience Rolle and Pärt both eagerly invite us into — and query our basic suppositions about sonorous ontology, the answer is no longer so clear.[26] Sounding voices no longer depend on vocal cords to enter into vibratory relationship with listening auditors; sound waves no longer mark the absolute limit of sound and its experience.

This is especially so when the gracious receipt of angelic harmony profoundly transforms the foundations of sonorous experience, whether we take that harmony as an article of faith or a complex cultural formation. Indeed, if we allow ourselves entry into a Rollean framework before asking these questions, what appears outwardly silent becomes full of inward sound; conversely, plenitude of outward sound, even when that sound is the liturgical celebration of Mass, silences the true music of the soul. Rolle "praises God in song, and yet in silence, yielding praises not to men's ears but in God's sight with a marvelous sweetness":[27] His language makes uncertain the distinction between song and silence as he invites us to reflect on the relative value of sound in the ear, in the imagination, and in the soul. This sonorous superposition, this quantum state of silent sound corresponds to the paradox of mystical communion with the divine, inviting us to go deeper into silence in order to hear our soul vibrating in the divine presence. What's more, it should hardly be unfamiliar to those versed in Pärt's pronouncements on the role of silence in his own work, to which I now turn.

Hearing Medieval: Desire, Script, and Silence in Arvo Pärt

Much has been written about the use and status of silence in Pärt's music.[28] Pärt himself has pronounced on the subject: "Silence must be approached with a feeling of awe," he says. "When we speak about silence, we must keep in mind that it has two different wings, so to speak. Silence can be both that which is outside of us and that which is inside a person. The silence of our soul, which isn't even affected by external distractions, is actually more crucial but more difficult to achieve."[29] The Rollean echoes are noteworthy: double angelic wings, a binarized internal/external framing of perceptual experience, the creative paradox of sound-in-silence. These convergences, interesting though they may be, are of dubious import, and this is to the point. Because their heritages are so disjunct, we can put aside questions of how Rolle influenced Pärt — a scholarly wild goose chase, to be sure — and instead focus on how the structural force a historical Rolle unleashes finds purchase in Pärt's music, how Rolle invites us to listen to Pärt differently, how medieval approaches to similar sound experiences might expand our hearing and invite us into productive new orientations to what we hear today.

Pärt asks us to listen for the silence in his music, to enter through our hearing into the same superposition of sound and silence that we see in Rolle's writings, an impossible hearing rooted in mysticism's well-worn rhetorical tropes of paradox and ineffability. Pärt's discourse about his own music valorizes this impossible hearing situation — the "silence of our soul" that his music implicitly promises "is actually more crucial but more difficult to achieve," a gauntlet thrown at our feet — and his music's frequent use of rich, episodic silence as compositional material builds pregnant meaning around the absence of sound. In a post-Cagean musical world, we of course recognize the absence of sound as a fiction. Yet Pärt's words and music both insist, in their own ways, on the reality of this pure silence, in part by moving it from outward perceptibility to the unverifiable closet of the soul, and they insist on this silence's availability through his music, so long as we have learned how to listen properly. We are thus encouraged to strive to hear in Pärt's music what *he* hears in his music, so that we might ourselves encounter the mysteries his music (and the discourse surrounding his music) promises — mysteries haloed with the aroma of authenticity, of the divine, and of the medieval.

To understand this desiring dynamic, we can profitably recall the mechanism of Rollean textuality, the way Rolle's *Melos amoris* scripts a rhythmic verbal sonority for our bodily ears that gestures toward harmonious angelic sonority without ever being able to deliver elevated spiritual audition directly. Pärt can be said to extend a similar promise, to generate a similar longing, for this pure, mystical silence his music can never supply. Seen from a slightly different angle, at the same time that the *Melos amoris* valorizes a spiritual audition it cannot provide, it does offer an immediate audition that is delightful in its own regard, a virtuosic display of verbal skill that can be admired and appreciated for its own excellence, even while it sustains a constitutive relation to the ineffable. Whether readers and auditors of the *Melos* attend to the text's virtuosity, to its spiritual referentiality, or both depends on the styles of hearing they bring to their reading. A text like the *Melos* thus sculpts the interface between external and internal sounds and silences into numerous possible configurations, each providing access to distinctive forms of sonorous pleasure, desire, and knowledge, some of them mutually compatible, others not. We might consequently understand the diversity of manners of listening to Pärt's music — popular, scholarly, cinematographic, affective, ambient, aesthetic, devotional, mystical — not as misapprehensions or impoverished auditions of that music but rather as hearing experiences that that music and the visual, cultural, and verbal discourses surrounding it carve out and invite. This invites us as critics to expand our parameters for what proper listening to Pärt's music is. Few listeners come to Pärt's music as hesychasts, yet many are capable of hearing and making meaningful the spacious silences Pärt's music creates. If we want to understand how Pärt's music achieves this, we need to think beyond the notes of his scores and consider the fuller perceptual, cultural, and spiritual frameworks for hearing that constitute the diversity of listeners' experiences.

This is not to say that Pärt's scores should be ignored. To the contrary, careful examination of Pärt's scores reveal the surprising ways these documents shape our encounter with the music they record, and medieval manuscript culture can help explain how that shaping work unfolds. To see how this might

Figure 1. First page of *Sieben Magnificat-Antiphonen*. Arvo Pärt "7 Magnificat-Antiphonen | für gemischten Chor a cappella" © Copyright 1990 by Universal Edition A.G., Wien/UE19098.

be, I turn to Pärt's setting of the seven O Antiphons, liturgical pieces that in their original Latin have framed the Magnificat during the last seven days of Advent since at least the seventh century. Figure 1 reproduces the first page of the Universal Editions score for Pärt's *Sieben Magnificat-Antiphonen* as found in the holdings of the Columbia University and the New York Public Libraries,

Figure 2. Detail from "O Weisheit." Arvo Pärt "O Weisheit" from "Collected Choral Works" © Copyright 2008 by Universal Edition A.G., Wien/UE33880.

the only two library copies of the score available in New York City. Though a neatly edited and typeset edition of *Sieben Magnificat-Antiphonen* can be purchased from the Universal Editions website, it is not inconsequential that performers and researchers seeking publicly archived scores of Pärt's composition can only find it in manuscript in as major an intellectual center as New York City. We can still see the mark of the composer's living hand on these scores, the *ductus* of his pen as it leaves a trail of ink on the page. In a digital cultural landscape where encountering another's handwriting is an increasingly rare occurrence, reading, interpreting, and performing Pärt's own marks from the photocopied page carries with it a Benjaminian aura of authenticity that the composer's mystical-medieval persona only serves to bolster.[30] This metonymic encounter with the present-absent composer must shape performers' relationship to the music, itself predicated on plays of sonorous presence and absent silence, which must in turn impact the performance they deliver to an audience — and this in dialogue with the matrix of expectations, desires, knowledges, and beliefs performers and audiences bring to the event of music.

A small headnote at the top right corner of this manuscript score (Figure 2) indicates that Pärt composed the work in 1988, the year written out in full, and revised it in May 1991, abbreviated and squeezed in under a demarcating virgule. The score we read presumably represents the 1991 revision, but this is not a clean copy: The process of revision and correction is still visible on the page, a situation familiar to those who work with medieval manuscripts. In the last system of the third antiphon, "O Sproß aus Isais Wurzel," for example

Figure 3. Detail from "O Sproß aus Isais Wurzel." Arvo Pärt "O Sproß aus Isais Wurzel" from "Collected Choral Works" © Copyright 2008 by Universal Edition A.G., Wien/UE33880.

(Figure 3), the original 1988 version of the *Sieben Magnificat-Antiphonen* lingers under erasure on the score, inaudible but refusing to be entirely silenced. Accounting for the revision history of *Sieben Magnificat-Antiphonen* grows even more complicated when we consult its listing in the Arvo Pärt Center's catalogue of Pärt's texted compositions, *In Principio* (Figure 4). There, it is listed under a variant title, *O-Antiphonen*, as a work scored for eight cellos, commissioned by Cello8ctet Amsterdam for premiere in 2008. No mention is made of the work's alternative title or its choral genesis two decades prior. Moreover, the Universal Editions website offers scores for purchase under both titles, identical in all respects but orchestration.

This encounter with Pärt's scores raises a range of questions. Where in Pärt's opus do we locate his setting of these seven liturgical antiphons? Should we call this work *Sieben Magnificat-Antiphonen* or *O-Antiphonen*; are we listening to different works when we hear choir and cello octet perform the same score? In which version, in which revision does Pärt's music reside? Questions like these presume a determinate musical object, a fixable, copyrightable thing subject to the economic and property logics of late capitalism. From a medieval perspective, however, these questions resolve with relative ease: In a culture where texts are copied by hand, human error and scribal fancy obviate any notion of textual stability, much to the consternation of modern editors seeking a reliably correct work for print. Theorists like Paul Zumthor and Bernard Cerquiglini have helped reveal how mobile the medieval text can be, how resistant to the much-debated "work concept," which seems ill-suited to Pärt's compositional practices.[31] Pärt's scores ask us to query how his music navigates notions of authorship, musical ontology, material and intellectual property, and the work of art in this our age of digital reproduction; the Middle Ages provides us with useful conceptual and theoretical tools to help in that endeavor.

Comparable mobilities, resistances, and ambiguities appear in Pärt's music

O-ANTIPHONEN

2008
for 8 violoncellos
Commissioned by Cello8ctet Amsterdam (former Cello Octet
Conjunto Ibérico).
Dedicated to Conjunto Ibérico.
Premiere: 21/10/2008 Amsterdam, Netherlands
Cello8ctet Amsterdam

Figure 4. Listing for O-Antiphonen, in Hedi Rosma, et al., eds., In Principio: The Word in Arvo
Pärt's Music (Laulasmaa, Estonia: Arvo Pärt Centre, 2014), 232. Courtesy of the Arvo Pärt Centre.

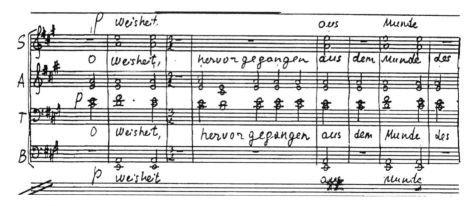

Figure 5. First system of "O Weisheit." Note text underlay for each voice part and fragmentation of text in soprano and bass. Arvo Pärt "O Weisheit" from "Collected Choral Works" © Copyright 2008 by Universal Edition A.G., Wien/UE33880.

when we turn to his manner of text setting, a topic that has received substantial treatment as a compositional technique but little attention as an expressive poetics.[32] In the first of the *Sieben Magnificat-Antiphonen*, "O Weisheit" (Figure 5), each voice part's text setting has been written out, M-voice in the tenor, T-voices building an A major triad in the other three parts. While the altos match the tenors word for word, the soprano and bass parts are populated with rests that cancel out more than half the words. A similar textual cancellation seems to occur in the next antiphon, "O Adonai" (Figure 6). Here, a single line of text appears, shared between tenor and bass parts as they trade an open fifth drone on F# and C# for almost the full length of the piece. While one voice sounds the drone, the other enunciates the text. In both examples, though by different scribal means, the relationships between words, voices, and music are made beguilingly ambiguous. Those moments when soprano and bass fall silent in "O Weisheit" are far from empty: How does their silence continue to speak the text? The vowel on which an ensemble ought to sing the drones in "O Adonai" is left unspecified, not unlike the untexted tenors of fourteenth-century motets: Why does Pärt leave this as a problem embedded in the score for the ensemble to solve? Seen from one angle, these gaps and ambiguities echo the fragmentation and uncertainty that so often characterize mystical reportage, limiting our access to the ineffable experience of the mystic who speaks to us from the other side, who writes "postcards from the edge," to borrow Vincent Gillespie's apt turn of phrase.[33] Seen from another angle, these gaps and ambiguities speak to Pärt's sense of shared artistic agency and make more flexible the relationship between composer and performer, words and music, language and voice.

Figure 6. First system of "O Adonai." Note shared text across all voice parts, including tied whole-note drones spanning multiple words. Arvo Pärt "O Adonai" from "Collected Choral Works" © Copyright 2008 by Universal Edition A.G., Wien/UE33880.

We may push the question one step further: In what way can a cello octet be said to perform the words of the seven O Antiphons? How do musical instruments unable to utter language bring into the world those words — liturgical words, sacred words — that Pärt's music ascribes to them? This is a question not only for the *Antiphonen* but for all the compositions in Pärt's catalogue built from texts but scored for instruments. Many of these compositions print the text from which they were generated in the score itself, accompanied by the prefatory note that "all text parameters (number of syllables, accentuations, punctuation, etc.) were determining factors in this composition. In order to illustrate the phrasing logic resulting from the text for musicians, it has been written underneath the score." We do well to apply healthy suspicion to this note's utilitarian thrust — it seems likely that the presence of Church Slavonic on the pages of *Trisagion* (Figure 7) does much more than offer phrasing guidance for string players, not least because string players who can read, much less understand, Church Slavonic are likely in the minority.

How do the Cyrillic syllables scattered underneath *Trisagion*'s systems affect a string orchestra's performance such that those words sound forth in their playing? How does Vulgate Psalm 112 sound forth in *Psalom*, when the Universal Editions score neglects to provide the biblical text on which it is based?[34] The skeptic's answer would be, of course, that it doesn't, but Pärt tells us in no uncertain terms that it does — "Sound is my word," "The words write my music"[35] — and while we may not share the articles of faith that underlie such claims, we do well to examine them for how they shape and frame our experience of performing and hearing Pärt's music. From a theological and devotional perspective, the words of scripture and prayer recall and participate in the Word of God who is Christ, who bridges the material and spiritual, whose

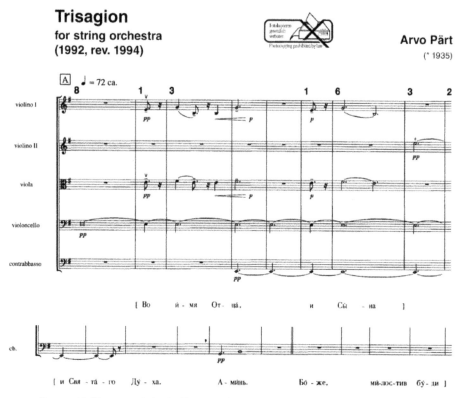

Figure 7. Cyrillic text underlay for *Trisagion*. Arvo Pärt "Trisagion | [Introductory Prayers]|für Streichorchester" © Copyright 1996 by Universal Edition A.G., Wien/UE31265.

name is best prayed in silence.[36] Rolle, an early proponent of devotion to the Holy Name, would surely find this outlook amenable. Pärt's tintinnabuli style likewise emulates the Christological union by knitting the M-voice's time-bound, errant melody to the triadic stability of T-voices that seems to stretch into eternity.

Still, the printed fact of these Cyrillic words, foreign in their script and medieval in their sacred idiom, evokes the arcane knowledge of an ancient religious tradition whose mysteries we catch only glancingly. The otherwise inaudible medieval breaks the surface of the score to leave its inscrutable mark. More basically, the unspoken words and reassuring prefatory note in Pärt's orchestral scores serve to certify that a method drives his compositions, a method we need not fully understand in order to hear. If Pärt's works solicit our desire for a spiritual silence his music cannot provide, or for a sacred text his instrumentation cannot pronounce, his compositions do make readily audible their algorithmic method, a method in whose consequent style we delight. Indeed, this is what perhaps is most striking about Pärt's music: that so deceptively simple a compositional method could produce a style broad audiences widely account beautiful.

Were we to ask medieval music theorists, they likely would attribute this beauty to *musica*: not sounding music per se but the rational proportionality of number that underlies music, determining the consonance of intervals, the equilibrium of human being, and the order of the cosmos.[37] Pärt might well concur — 1+1=1, after all — though for him, *musica* and *logos* appear to converge in ways that medieval Western theorists did not find so interesting. Yet whether Pärt concurs or not — indeed, the entire question of what Pärt believes — is largely beside the point when we seek to better understand how Pärt's music renders meaningful, moving experiences for audiences, performers, and even scholars. Instead of privileging the composer's intentions, his belief system, or his cultural poetics, we might squarely locate the beauty of Pärt's music in its present sounding, in its echoing silent spaces, in its orderly arrangement of materials — in its structures, yes, but also in the human bodies that make those structures sound, and in those more abstruse, pervasive structures his music receives and sets into vibratory motion. Richard Rolle, and the Middle Ages more broadly, stand to help us enter into this alternative audition with greater acuity, if we allow them. By listening with medieval ears, we can hear Pärt's music not just in our present but also in our past, a past of our own invention that yet exerts energetic force across historical distances, a dislocating medieval past that can teach us how to hear anew.

Notes

1. See Leopold Brauneiss, "Arvo Pärt's Tintinnabuli-Style: Contemporary Music toward a New Middle Ages?" in *Postmodern Medievalisms*, Studies in Medievalism 13 (Woodbridge: Boydell & Brewer, 2005), 27–34.

2. For varied engagements with this simultaneously othering and nostalgic dynamic, see Thomas Hahn, ed., "The Medievalism of Nostalgia," special issue, *postmedieval: a journal of medieval cultural studies* 2, no. 2 (Summer 2011); and Nicholas Watson, "Desire for the Past," *Studies in the Age of Chaucer* 21 (1999): 59–97.

3. Jeremy Begbie partially attributes the "immense popularity" of Pärt's music to its creation of "a cool sonic cathedral in a hot, rushed, and overcrowded culture." Jeremy S. Begbie, *Resounding Truth: Christian Wisdom in the World of Music* (Grand Rapids, MI: Baker Academic, 2007), 261.

4. Wolfgang Sandner, CD liner notes for Arvo Pärt, *Tabula rasa*, trans. Anne Cattaneo (ECM New Series 1275, 1984), 27.

5. Enzo Restagno et al., *Arvo Pärt in Conversation*, trans. Robert Crow (Cham-

paign, IL: Dalkey Archive, 2012), 62. Peter C. Bouteneff explores this monkish persona at greater length in *Arvo Pärt: Out of Silence* (Yonkers, NY: SVS, 2015), 43–47.

6. Richard Taruskin, "Sacred Entertainments," *Cambridge Opera Journal* 15 (2003): 109–26; and Robin Holloway, "Beware the Pitfalls of Sincerity," in *Essays and Diversions, 1963–2003* (Brinkworth: Claridge, 2003), 294–96. Robert Sholl comments on this strand of critique in "Arvo Pärt and Spirituality," in *The Cambridge Companion to Arvo Pärt*, ed. Andrew Shenton (Cambridge: Cambridge University Press, 2012), 148–50. In the same volume, Benjamin Skipp, "The Minimalism of Arvo Pärt: An 'Antidote' to Modernism and Multiplicity?" reports Pärt's musical resistance to modernity approvingly, writing that "tintinnabuli, as with minimalism generally, is situated as a counter-modernist reaction" (165) and that "while the temptation to caricature Pärt as a hermit removed from modern life should be resisted, he does not conform to images of progress and egoistic glory which infuse numerous strains of modernity" (168).

7. For modern temporalities, see Gerhard Dohrn-van Rossum, *History of the Hour: Clocks and Modern Temporal Orders*, trans. Dunlap Thomas (Chicago: University of Chicago Press, 1996); and Reinhart Koselleck, *Futures Past: On the Semantics of Historical Time*, trans. Keith Tribe (Cambridge, MA: MIT Press, 1985). For medieval temporalities, see Carlo M. Cipolla, *Clocks and Culture: 1300–1700* (New York: Norton, 1977); Aron Gurevich, "What Is Time?" in *Categories of Medieval Culture*, trans. G. L. Campbell (London: Routledge, 1985), 93–152; Chris Humphrey and W. M. Ormrod, eds., *Time in the Medieval World* (Rochester, NY: York Medieval Press, 2001); Gerhard Jaritz and Gerson Moreno-Riano, eds., *Time and Eternity: The Medieval Discourse* (Turnhout: Brepols, 2003); Jacques Le Goff, *Time, Work, and Culture in the Middle Ages*, trans. Arthur Goldhammer (Chicago: University of Chicago Press, 1980).

8. The critique of teleological historicity owes much to scholarship inside and outside medieval studies that seeks to queer received notions of time and temporality. See, for example, Dipesh Chakrabarty, "Where Is the Now?" *Critical Inquiry* 30, no. 2 (Winter 2004): 458–62; Carolyn Dinshaw, *How Soon Is Now?: Medieval Texts, Amateur Readers, and the Queerness of Time* (Durham, NC: Duke University Press, 2012); Carla Freccero, *Queer/Early/Modern* (Durham, NC: Duke University Press, 2006); and Carolyn Dinshaw et al., "Theorizing Queer Temporalities: A Roundtable Discussion," *GLQ: A Journal of Lesbian and Gay Studies* 13, no. 2 (2007): 177–95. For the use of the Middle Ages by later historical moments for various, often troubling, cultural and political ends, see Andrew Albin et al., eds., *Whose Middle Ages?: Teachable Moments for an Ill-Used Past* (New York: Fordham University, 2019).

9. This mode of criticism is sometimes termed posthistorical, to differentiate it from the new historicist approaches that have dominated literary studies for the last two decades. See, for example, Elizabeth Scala and Sylvia Federico, eds., *The Post-Historical Middle Ages* (New York: Palgrave Macmillan, 2009); and Paul Strohm, "Historicity without Historicism?" *postmedieval: a journal of medieval cultural studies* 1 (2010): 380–91.

10. Shai Burstyn adapts Michael Baxandall's art historical scholarship on Renaissance visual culture to outline the concept of a "period ear" in "In Quest of the Period Ear," *Early Music* 25, no. 4 (November 1997): 692–701. Where Burstyn seeks to acquire this ear in order to listen to the music of the past with greater emic accuracy, I am interested in bracketing the question of accuracy and exploring more contingently how historical listening practices might disrupt and reorient received habits of contemporary audition.

11. Rolle's conversion experience was infamous enough to be recorded in his votive office, prepared thirty or so years after his death. Returning to his hometown after leaving Oxford without his baccalaureate, Rolle instructs his sister to bring two of her robes and their father's rain hood to the edge of a forest. Garments in hand, he tailors them on the spot, strips naked, and dons them in imitation of a hermit's habit. Rolle's scandalized sibling flees the scene, crying out, "My brother's gone crazy!" (Frater meus insanit) — not the last time Rolle would hear the accusation — and the young man in turn flees family and friends to a new religious life. See Reginald Maxwell Woolley, ed., *The Officium and Miracula of Richard Rolle of Hampole* (New York: Macmillan, 1919), 24.

12. The most influential critical account of Rolle's life and works remains Nicholas Watson, *Richard Rolle and the Invention of Authority* (Cambridge: Cambridge University Press, 1991). For a concise biography, see Jonathan Hughes, "Rolle, Richard (1305x10–1349)," *Oxford Dictionary of National Biography* (Oxford, 2008), http://www.oxforddnb.com/view/article/24024. Further contexts relevant to the present chapter may also be found in my *"The Melody of Love:* Ten Ways In," in *Richard Rolle's Melody of Love: A Study and Translation with Manuscript and Musical Contexts*, Studies and Texts 212 (Toronto: Pontifical Institute of Mediaeval Studies, 2018), 1–133.

13. Walter Hilton's late fourteenth-century epistle "Of Angels' Song" walks a delicate line between maintaining Rolle's orthodoxy, cautioning readers against presuming they have achieved Rollean *canor*, and holding out the possibility that in rare cases *canor* can authentically be achieved. Thomas Bassett's early fifteenth-century defense of Rollean mysticism, *Defensorium contra oblectratores*, and the mystical writings of his contemporaries Richard Methley and John Norton witness the continued growth of interest in and wariness toward the hermit's sensorial mysticism, which more austere critics viewed to be sensorial excesses.

14. For comparison's sake, Geoffrey Chaucer's *Canterbury Tales* survives in eighty-three manuscripts, *Sir Gawain and the Green Knight* in just one.

15. The following account of *canor* and Rollean textuality receives fuller treatment in Andrew Albin, "Listening for *Canor* in Richard Rolle's *Melos amoris*," in *Voice and Voicelessness in Medieval Europe*, ed. Irit Ruth Kleiman (New York, 2015), 177–97.

16. "Dum enim in eadem capella sederem, et in nocte ante cenam psalmos prout potui decantarem, quasi tinnitum psallencium uel pocius canencium supra me ascultaui. Cumque celestibus eciam orando toto desiderio intendem, nescio quo-

modo mox in me concentum canorum sensi, et delectabilissimam armoniam celicus excepi, mecum manentem in mente. Nam cogitacio mea continuo in carmen canorum commutabatur, et quasi odas habui meditando, et eciam oracionibus ipsis et psalmodia eundem sonum edidi. Deinceps usque ad canendum que prius dixeram, pre affluencia suauitatis interne prorupi, occulte quidem, quia tantummodo coram Conditore meo." Richard Rolle, *The Incendium Amoris of Richard Rolle of Hampole*, ed. Margaret Deanesly (New York: University of Manchester, 1915), 189–90.

17. Katherine Zieman argues this point at greater length in "The Perils of *Canor*: Mystical Authority, Alliteration, and Extragrammatical Meaning in Rolle, the *Cloud*-Author, and Hilton," *Yearbook of Langland Studies* 22 (2008): 131–63, esp. 138–40.

18. Later in the *Incendium*, Rolle admits that "I have sought to flee the outwards songs they repeat so often in church, and the polyphonic melodies you hear while standing there" (cantica exteriora que in ecclesii consueta sunt frequentari, organica quoque modulamina que ab astantibus audiuntur fugere curaui), explaining that outward music interferes with his ability to inwardly hear the heavenly song that always accompanies him. Rolle, *Incendium*, 233. Parallel justification for his absence at Mass appears in chapters 45 and 46 of the *Melos amoris*. In chapter 6 of the *Melos*, the hesychastic qualities of Rolle's inward singing grow somewhat more pronounced when he reveals that an influx of grace can sustain his participation in angelic song even when surrounded by the cacophony of the world: "These arias evaporate among others unless an amorous impulse assists me when I'm also engulfed in elation — I'm removed then from the rabble's racket and idiots' ribaldry" (Illud autem inter homines non habeo nisi propter impetum amoris et quando eciam ardore affluo, ammoto strepitu circumastancium et turpitudine stultorum). Translation in Albin, *Melody of Love*, 149; Latin in Richard Rolle, *The Melos Amoris of Richard Rolle of Hampole*, ed. E. J. F. Arnould (Oxford, 1957), 17.

19. ". . . quis in cantico intus et extra vitaliter inclusus. Sensus sic subsistunt in intimis intenti et unum tam arte et avide astringunt, universis iam aliis penitus exclusis, quod fluunt nequaquam nec exeunt ad ista." Rolle, *Melos*, 136.

20. "Mundi quippe amatores scire possunt uerba uel carmina nostrarum cancionum, non autem cantica nostrorum carminum; quia uerba legunt, sed notam et tonum ac suauitatem odarum addiscere non possunt." Rolle, *Incendium*, 278.

21. Rolle, *Melos*, 52–53. Translation in Albin, *Melody of Love*, 187–88.

22. This argument is developed in full in Albin, "Listening for *Canor*"; and Albin, *Richard Rolle's Melody of Love*, 27–30. Bouteneff's account of the Orthodox theology underpinning Pärt's compositional ethos in his *Arvo Pärt: Out of Silence* suggests numerous points of conversation between the Estonian composer's and the medieval English mystic's outlooks, in particular with respect to the active presence of the divine in music's structure, composition, and performance.

23. The most comprehensive study of reading practices in late medieval England is Joyce Coleman, *Public Reading and the Reading Public in Late Medieval England and France* (New York: Cambridge University Press, 1996). Katherine Zieman, *Sing-*

ing the New Song: Literacy and Liturgy in Late Medieval England (Philadelphia: University of Pennsylvania Press, 2008), investigates forms of literacy during the period through the lens of liturgical worship.

24. For more on medieval experiences of reading from parchment, see Sarah Kay, *Animal Skins and the Reading Self in Medieval Latin and French Bestiaries* (Chicago: University of Chicago Press, 2017).

25. Derrida's famous challenge to the traditionally understood metonymic vocality of writing in "Signature Event Context," in *Limited Inc.* (Evanston, IL: Northwestern University Press, 1988), 1–23, has prompted more sophisticated study of the complex imbrication of vocality, textuality, and presence, with respect to medieval manuscripts in particular. Jesse Gellrich offers an especially persuasive account of how orality, vocality, and textuality cooperate ideologically in late medieval textual culture; see "*Vox Literata*: On the Uses of Oral and Written Language in the Late Middle Ages," in *Discourse and Dominion in the Fourteenth Century: Oral Contexts of Writing in Philosophy, Politics, and Poetry* (Princeton, NJ: Princeton University Press, 1995), 3–36.

26. Sonorous ontology remains a live area of scholarly debate in the fields of deaf and sound studies. See Michele Friedner and Stefan Helmreich, "Sound Studies Meets Deaf Studies," *Senses & Society* 7, no. 1 (2012): 72–86; Rebecca Sanchez, "Deafness and Sound," in *Literature and Sound*, ed. Anna Snaith (Cambridge: Cambridge University Press, forthcoming). See also Steve Goodman, *Sonic Warfare: Sound, Affect, and the Ecology of Fear* (Cambridge, MA: MIT Press, 2010); Greg Hainge, *Noise Matters: Towards an Ontology of Noise* (New York: Bloomsbury Academic, 2013); and Brian Kane, "Sound Studies without Auditory Culture: A Critique of the Ontological Turn," *Sound Studies* 1, no. 1 (2015): 2–21.

27. "Laudat Deum in iubilo, sed in silencio; non ad aures hominum sed in conspectu Dei et ineffabili suauitate odas emittit, id est laudes." Rolle, *Incendium*, 238.

28. See, for example, Bouteneff, *Arvo Pärt*, esp. 85–137; Andreas Kähler, "Radiating from Silence: The Works of Arvo Pärt Seen through a Musician's Eyes," in *The Cambridge Companion to Arvo Pärt*, ed. Andrew Shenton (Cambridge: Cambridge University Press, 2012), 193–97; Diego Malquori, "Music beyond Silence: A Reading of Arvo Pärt's *Credo*," *Musical Times* 157, no. 1936 (Autumn 2016): 37–48.

29. Tom Huizinga, "The Silence and Awe of Arvo Pärt," National Public Radio, June 2, 2014, https://www.npr.org/sections/deceptivecadence/2014/06/02/316322238/the-silence-and-awe-of-arvo-p-rt.

30. Walter Benjamin, "The Work of Art in the Age of Mechanical Reproduction," in *Illuminations*, ed. Hannah Arendt, trans. Harry Zohn (New York: Schocken, 1968), 217–52.

31. Zumthor develops his theory of *mouvance*, the authorially decentered textual variability of medieval texts as they circulated in manuscripts, in *Essai de poétique médiévale* (Paris: Éditions du Seuil, 1972) and *La lettre et la voix: de la "littérature" médiévale* (Paris: Éditions du Seuil, 1987). Cerquiglini builds on Zumthor's writings to develop an approach to textual editing in *In Praise of the Variant: A Critical His-*

tory of Philology, trans. Betsy Wing (Baltimore, MD: Johns Hopkins University Press, 1999). The musicological literature on the work concept is vast. A few points of entry include Lydia Goehr, *The Imaginary Museum of Musical Works: An Essay in the Philosophy of Music* (Oxford: Clarendon, 1992); Michael Talbot, ed., *The Musical Work: Reality or Invention?* (Liverpool: Liverpool University Press, 2000); Harry White, "'If It's Baroque, Don't Fix It': Reflections on Lydia Goehr's 'Work-Concept' and the Historical Integrity of Musical Composition," *Acta Musicologica* 69, no. 1 (1997): 94–104.

32. Leopold Brauneiss explains tintinnabuli compositional technique in "Musical Archetypes: The Basic Elements of the Tintinnabuli Style," in *The Cambridge Companion to Arvo Pärt*, ed. Andrew Shenton (Cambridge: Cambridge University Press, 2012), 49–75.

33. "Postcards from the Edge: Interpreting the Ineffable in the Middle English Mystics," in *Interpretation Medieval and Modern: The J. A. W. Bennett Memorial Lectures: Perugia 1992*, ed. P. Boitani and A. Torti (Cambridge: Cambridge University Press, 1993), 137–65.

34. Bouteneff asks similar questions in *Arvo Pärt*, 75–76.

35. Hedi Rosma et al., eds., *In Principio: The Word in Arvo Pärt's Music* (Laulasmaa, Estonia: Arvo Pärt Centre, 2014), 5, 13.

36. Bouteneff, *Arvo Pärt*, 116–23, 139–76.

37. Boethius famously describes these three species of *musica* in *De institutione musica* 1.3. For a deeply considered account of how medieval music theory from Boethius onward amounted to much more than a matter of mathematics, see Andrew Hicks, *Composing the World: Harmony in the Medieval Platonic Cosmos* (Oxford: Oxford University Press, 2017).

10
The Piano and the Performing Body in the Music of Arvo Pärt

Phenomenological Perspectives

Maria Cizmic and Adriana Helbig

Introduction

In this chapter, we draw on literature in musicology and ethnomusicology regarding embodiment and phenomenology to consider the piano and Arvo Pärt. We examine references to the piano in existing literature on Pärt and articulate a perspective on the piano's role in his music. Pärt's compositional output for the piano is relatively small, but, as his primary instrument, the piano has played an important role in the development of his sonic world. Musicians' physical relationships with their instruments over time generate an archive of embodied knowledge and memories; bodily performance then becomes an important way that music creates meaning for composers and performers alike. Thus, this chapter draws on representations of Pärt composing at the piano as captured in cinema, analyzing the bodies of composer and instrument in dynamic engagement. It also features a close observational analysis of performing *Für Alina*, Pärt's inaugural "tintinnabuli" work from 1976, and discusses what a phenomenological experience of learning and performing this piece might tell us about the expressive nature of tintinnabuli music.

Sounding the Body

As many ethnomusicologists and musicologists have argued, the physical body is central to the production of music and musical meaning. Work in phenomenological ethnomusicology has brought us closer to understanding the interplay of music, structure, perceptual agency, and cultural context.[1] Phenomenological thought has helped us understand that there are multiple ways that people create, perform, and hear sounds. Our bodies and our senses shape our perceptions of what surrounds us and help guide our engagement with the world. We are, in turn, ourselves changed by our physical experiences. As Merleau-Ponty argues, our bodies are conduits of our physical experiences and help process the sensations that surround and shape us.[2] If we hold this premise to be true, we may argue, then, that the human body is the musician's most important tool — both for creating music and for understanding musical meaning.

The sense and sensations of musical gestures sit at the center of our project. Gestures, thoughts, feelings, and articulations function simultaneously in response to stimuli that come from within and outside the body. As the ethnomusicologist Matthew Rahaim reminds us, "there is much to learn from gesture, posture, and the physicality of vocal production that cannot be learned from sound alone."[3] The growing literature on music and gesture encourages us to consider "personally, historically, or culturally specific characters of bodily manner, perhaps approaching a notion of habitus."[4] In this chapter, co-written by an ethnomusicologist and a musicologist, we seek to bring these various yet interrelated discourses together. We share a grounding in phenomenology and a focus on bodies and instruments, gesture, and performance.

Bodies persist through the multiple and overlapping arenas we consider here — composition, history, cinema, music performance — and they intersect with various types of memories. In considering both composer and performer, we think of the memories that center on a piece of music as a set of intersecting but not wholly commensurate experiences. Pärt's own personal, musical memories create one context, which he carries forward into his compositions. Certain gestures and bodily stances that may be personally meaningful to him work their way into his music, only to be transformed through a composition's form, instrumentation, sound, text, and so on. On the other end, performers carry their own archive of physical gestures and musical memories, which get both activated and expanded when learning to play new music. A performer's work makes sense in relationship to past experiences and memories, and at the same time new music expands one's bodily knowledge. Because musical performance can activate various sets of memories (for composers, performers, and listeners), there will naturally arise a set of meanings that bear some relationship to one another at the same time that they diverge. In a notated musical idiom, the score sits at the crux of these intersecting sets of memories. A score can seem to be a stable entity, given its status as a text, but it is constantly activated anew depending upon the context and nature of performance and recording. In phenomenological terms, we can think of embodied knowledge and memories, which composers, performers, and audiences carry with them, as the habitus and context within which musical performance makes sense. With this in mind, this chapter first looks to biographical and cinematic sources to consider how Pärt's relationship with the piano has shaped his compositions, leading to a focus on the piano's pedal.

Within a general context of classical music training in the late twentieth and early twenty-first centuries, we wish to highlight the ways that musicians' embodied experiences with instruments serve as a meaningful frame when creating and performing music. The sociologist David Sudnow, in his closely observed discussion of learning to play jazz piano, describes the nuances of what it takes to learn to play an instrument and the negotiation of the instrument's physical spaces through bodily gesture, practice, and repetition. Although Sudnow does not explicitly consider the role of musical style or genre, he does lead us to conclude that extended practice of a particular musical idiom on a specific instrument leads to "incorporation." Sudnow uses this word to describe how people generally become physically familiar with technology and, in this case, to identify how musicians internalize the physical spaces of their instrument so that it feels as familiar as a part of one's own body.[5] Through training in a particular style, musicians know how to move their hands in space to play a chord, to leap to a note, to perform the gestures of musically idiomatic patterns. With this in mind, training in a particular musical tradition for composers and performers alike forms a habitus, a meaningful framework and context. The second half of this essay includes a discussion by Maria Cizmic of learning to play *Für Alina*, conveying one musician's thoughts on how Pärt's work activates expectations and bodily knowledge based in classical piano training in tonal repertoire but then reformulates these habits. *Für Alina* contains compositional elements redolent of tonality, but by shifting attention to the resonance and decay of sound, in part through the use of the pedal, Pärt asks a pianist to change ingrained bodily habits. Much of the reception of tintinnabuli has been wrapped up with the concept of simplicity. But, through this phenomenological turn, we attempt to recast the common understanding of Pärt's "simplicity" in terms of embodied habits and musical expressivity.

A Riposte to Simplicity

If we pay attention to what Pärt says in his conversations with the Italian musicologist Enzo Restagno, we can introduce some much-needed nuance into two ossified narratives that surround tintinnabuli: first, that tintinnabuli was born out of a period of silence; second, that simplicity rules. The period between 1968 and 1976 was certainly not literally silent for Pärt. He describes his exercises in "automatic" composition during this time — he would read a psalm, or look at a flock of birds in the sky, and then write a melody quickly without too much thought. Pärt filled copious compositional diaries trying to translate such nonmusical experiences into spontaneous melody.[6] When it came to *Für Alina*, Pärt chose one of these melodies at random, created a "small edifice" for it, and added a second melody using the B minor triad.[7] *Für Alina* has many of the defining attributes of Pärt's music for years to come: a reexamination of tonality that foregrounds diatonic dissonance, a process involving addition and subtraction, polyphony, and slow tempos coupled with widely spaced intervals that allow for overtone-rich reverberation. Pärt characterizes his initial reaction to the piece by saying that it took him a while to come to terms with his own creation; he tells Restagno that, in the wake of composition, *Für Alina* "seemed too simple to me, my ears were not yet used to taking a piece like this seriously."[8] Silence is not absolute; simplicity was not immediately an idealized state of musical being. Pärt's comments are all the more interesting because they reflect back on a time (in 1976) not long after he had told Ivalu Randalu in 1968 that there is "wisdom in reduction."[9]

Of course, much of what Pärt has been quoted as saying — like the comment to Randalu — has rejected the complexity of musical modernism and valorized simplicity. Perhaps no comment is as iconic as Pärt's words included by Wolfgang Sandner in the ECM *Tabula rasa* liner notes: "It is enough when a single note is beautifully played."[10] In response to these kinds of statements, Pärt reception (be it scholarly or journalistic) has fallen roughly into two categories. One group finds meaning in simplicity and connects it to interiority, reflection, spirituality, depth, profundity, and a sense of repose as an antidote to our "harried times." Laura Dolp characterizes this reception as a form of "social minimalism" — musical simplicity relates to a set of idealized lifestyle goals.[11] For another group, simplicity betrays a lack of depth, meaning, ideas, and musical value. As a recent example, the composer James MacMillan and the theologian Jeremy Begbie criticize tintinnabuli as simple and express a view that endows musical complexity with value, seriousness, and potential for responsible social critique (a classic modernist position, in the vein of Adorno).[12] This push and pull between the values and associations around the terms "simplicity" and "complexity" sometimes seem to hold no exit.

One way out of this impasse, or to complicate it at the very least, is to return to Pärt's words. Here is one passage from 1999: "These days I want to make something 'palatable.' This has nothing to do with accommodating every musical taste within an audience. There has to be, however, a balance between the human perceptive faculty and the musical presentation. All important

things in life are simple."[13] And here is a more recent comment from his conversations with Restagno:

> A logical structure can work as a foundation when it is based on
> simple, comprehensible concepts. If things become exaggeratedly
> complicated, as if [sic] often the case with much contemporary music,
> people can no longer follow the musical thoughts of the composer . . .
>   I refer simply to the functional technical structuring of the piece,
> whose construction should be perceivable. . . . What I mean is that
> one should be able to move within clearly outlined borders of compre-
> hensibility.[14]

On the one hand, Pärt's valorization of simplicity and rejection of complexity and modernism are on full display in both passages. But rather than treat
simplicity and complexity as absolute values in and of themselves, Pärt repeatedly discusses simplicity and complexity in terms of a bigger issue: musical
expressivity and comprehensibility. Pärt rejects "contemporary music" (a.k.a.
"modernist music") for disregarding accessibility, and he seems to understand
comprehensibility as an issue for both composers' and audiences' perception
of music. His famous emphasis on simplicity, present in these quotations, is
consistently linked to the issue of expression and accessibility.

Because Pärt emigrated from Soviet Estonia to Austria and then West Germany in 1980, his early tintinnabuli works perhaps uniquely belong to both
the late socialist context (where they were composed) and the late capitalist
context (where they were recorded at ECM). During this time, musical expressivity and comprehensibility were significant to both locations, although
the stakes of accessibility were quite different depending on geography. In the
West, artistic accessibility was wrapped up with market forces; in the East,
accessibility was wrapped up with socialist realism. Peter Schmelz traces this
generation's arc from abstraction to mimesis, from twelve-tone modernism
to postmodern reformulations of tonality, and demonstrates that the Soviet
musical establishment favored compositions that had *dramaturgia*.[15] At the
same time that socialist realism demanded comprehensibility, we should not
reduce the significance of accessible musical idioms only to the meanings
made possible by official culture. Certainly Pärt was not alone in thinking
about expressivity and comprehensibility, a concern shared by many of his
generation across Central and Eastern Europe. We can think about the issue
of accessibility as a feature of the late socialist musical landscape as composers
tried to find alternatives to both socialist realism and modernism, at the same
time that their counterparts in Western Europe and the United States created
parallel ways out of modernism.

We only suggest here what could be a fuller historical argument; our focus at present is on the phenomenological attention to music-making bodies and what such an approach might suggest about the expressivity, comprehensibility, and accessibility of tintinnabuli music. For now, in order to think about musical expressivity, our aim is to set aside the discourse of simplicity that encircles tintinnabuli. Existential phenomenologists deem perception as a significant location of historically and culturally situated meanings; they observe and describe experience in an effort to peel away habits and assumptions that usually encase experience. In what follows, we perform a version of what Husserl calls *epoché*, a self-conscious setting aside (or bracketing) of a preexisting idea about experience. In *Listening and Voice*, Don Ihde characterizes epoché as itself a hermeneutic act.[16] And so, we address and then attempt to set aside the discourse of simplicity in order to focus on instruments and bodies. As Le Guin's work on Boccherini indicates, composers' physical relationships to instruments imprint upon their compositions. Biographical and cinematic evidence provides some insight into Pärt's relationship with the piano, one component of his musical habitus. By bracketing simplicity, we aim to consider more carefully how Pärt's privileging of high and low registers and his use of the sustain pedal create an expressive musical realm focused on resonance, sustain, and the decay of sound. The visual and sonic strategies of documentary films about Pärt draw our attention to the composer's relationship to the piano.

Pärt and Piano

Dorian Supin's documentary *And Then Came the Evening and the Morning* (1990) opens with a view of Pärt's head. As the camera moves away, we come to realize that this is his own reflection on the piano lid. Our intuition tells us that he is standing, body hunched over the instrument. He adjusts his head, closes his eyes, thinking, his facial expressions indicating that he is hearing sounds, listening. He shifts his position, resting his forehead now on the piano's edge, the cross around his neck moving in rhythm with his body. We see his hands move purposefully across the keyboard. *Cantus in Memory of Benjamin Britten* cascades in the soundtrack, as if the piece, whose score does not include a piano, emerges, like all the others, from the piano. This opening sequence shows us that Pärt's physicality is central to his compositional process, and through this juxtaposition of image and sound Supin places Pärt's piano-playing body as the source of all his compositions. His posture, his touch, and his patterns of breathing are ones a piano player will recognize. The body's positioning at the piano shapes the timbre, volume, and the overall energy of the music produced. In the video, Pärt appears to be resting on the piano, yet a musician exerts his greatest energy when leaning into the instrument. The left hand contorts, almost needlessly. These shapes indicate purpose, intent, and drive, rendering the resulting sounds deceptively simplistic. Yet these sounds are the result of intense emotions felt just before the fingers touch the keys. Even the slightest changes in feeling, instantly traveling between the mind and heart down to the fingertips, will change the sound of the note.

In *Beethoven the Pianist* (2010), Tilman Skowroneck points to two historical opinions about how musicians are inspired by their instruments. One school of thought views a particular style as a consequence of the instrument's properties and believes that exposure to a single type of instrument inevitably influences the player in a predictable way: Composers write for the way their instrument sounds. The second school of thought interprets a particular style of composing as a compensation for the shortcomings of the instrument or, ideally, trusts the capability of the player to make the instrument sound especially well, "pointing into the future."[17] Recent work on timbre has dealt with similar terrain; Emily Dolan takes a materialist position and traces out the history of the ideas of tone and timbre in order to demonstrate how Haydn composed his pieces for the physical capabilities of particular performers and instruments. She positions her argument as a correction to the ways that orchestration became marginalized in musical discourse of the twentieth century.[18] Pärt has frequently asserted that his compositions exist independently of the instruments that play them. Despite his neo-Platonic stance, our approach here is more materialist in orientation and looks to the physical capabilities of the piano to shed light on Pärt's music.

Human beings constitute their realities in relation to what they know or believe to be true. We process our experiences through our knowledge based on what we encounter through our senses. We compare these experiences with past thoughts, feelings, and actions, the memories of which we draw on to shape and process our present and future actions and beliefs. The philosopher Martin Heidegger articulates this point when it comes to sound. In *Being and Time*, Heidegger argues that it is difficult to comprehend sound without context. He states that the human being is unable to "hear" a "pure noise." Rather, what we hear "initially" are "the creaking wagon, the motorcycle . . . the column on the march, the north wind, the woodpecker tapping, the fire crackling."[19] We first perceive the object and then comprehend the sound in relation to that object. The human body seeks patterns and the ability to make sense of sounds in order to be able to communicate and comprehend. Heidegger's words seem to suggest a relationship between context and the perception of sound: If listeners approach Pärt's music with a firm narrative of simplicity and "social minimalism," then they listen for attributes that correspond to that context. Here we attempt to shift the context to embodiment and expressivity. Performances also provide ever new and additional contexts and interpretations of existing works. If we believe that every iteration of a musical composition is a new incarnation and perceived differently by us and others based on our state of being in the world, where do we root our understandings of the ways humans process sonic and musical meaning?

The memory of an instrument in our previous experiences causes us to relate to the instrument as potential sound object in a particular way. In other words, whatever instruments we encounter, our imaginations draw upon those past embodied memories to anticipate the potential of sound. Paul Hillier shares a story about Pärt's childhood piano.[20] Hillier explains that there was no music in Pärt's life until his family rented a house that had a grand piano. Beginning piano lessons at the age of eight, Pärt quickly learned to supplement the meager beginner's repertoire with his own improvisations. From the time he was twelve, he was able to notate his compositions. Hillier asserts that the piano, in poor condition with most of its middle register broken, had a profound influence on how Pärt perceived music. He contrasted the bass and the high treble registers and, in his imagination, filled in what was to come in between. Hillier reasons that this love for a wide range of pitches became an evident trademark for Pärt.[21]

Building on Hillier's observation about registers on Pärt's childhood piano, it seems that many answers to Pärt's musical output are, in fact, tied to his relationship with the piano's right pedal. The piano's sustain pedal lifts the dampers from all strings, sustaining all played notes and adding resonance to the sound of the instrument from sympathetic vibrations.[22] The low and high sustained pitches so iconic in Pärt's style grow out of the piano's capabilities. The middle registers on the piano are limited by the instrument's physical structure as regards their capabilities to sustain pitch. The sustain pedal has been found to increase the decay time of partials in the middle range of the keyboard, but this effect is not observed in the case of the bass and treble tones.[23] In other words, the pedal does not sustain all notes for a similar duration but rather privileges the lower and upper registers.

Gunter Atteln's 2015 film *The Lost Paradise: Arvo Pärt and Robert Wilson* captures Pärt's use of the sustain pedal during the compositional process. For only a few seconds in the film, we encounter Pärt, seated at the upright piano in the main room of the Arvo Pärt Center. Playing a few notes with his right hand, he brings his left hand to his lips, deep in thought. The sound is present but also slips from our hearing abilities as it fades, taking up space as silence, held by the pedal. The instrument, an upright, is slightly out of tune. It is significantly less expensive than the instruments upon which his music is now performed at concerts and in recordings. The sounds that emanate from an upright differ greatly from that of a grand piano, varying significantly in terms of what the instrument itself can do. The instrument offers the performer a range of sonic possibilities and challenges the player to engage with its physicality to produce the sounds the performer imagines.

The sustain pedal is perhaps the most complex part of the piano's mechanism. Best compared to the clutch on a manual transmission, its proper use cannot be taught through words alone; it must be felt. Thus, we come full circle to earlier arguments about the central role of the body in Pärt's music. Yet if the piano is as central to Pärt's compositional process as these documentations indicate, then we must consider a bifurcated temporality that comes into play with the piano. The piano allows for a consciousness of the present and the future. Unlike any other instrument, the piano, with its sustain pedal, embodies a type of temporal displacement in its construction. A note can be played and sustained through the pedal. But a pianist will play the note differently if choosing ahead of time to sustain the note with the pedal. Thus, these nuances of piano performance practice become ever more crucial when considering intent. And in watching Pärt at the piano, we can glean insights into the complexities of touch that shape the ways his music sounds and resounds. Together, the sustain pedal and the privileging of high and low registers emphasize the resonating capacity of the piano and shift performers' and listeners' attention to the temporal sustain and decay of sound. As Hillier and the documentary images suggest, these musical attributes are grounded in the piano at the same time that they persist in Pärt's writing for other instruments and use of the recording studio.

Für Alina, 1976

A number of sources suggest an attention to phenomenology and embodiment as a productive avenue for understanding Pärt's music. In addition to Hermann Sabbe's phenomenological description of *Tabula rasa* and Robert Sholl's suggestion that Pärt's music encourages a phenomenological attitude in a listener as part of his disenchantment with modernism,[24] Nora Pärt explains that tintinnabuli music is "more for the ear than the intellect."[25] Her observation draws our attention to the sensory experience of music, which we have sought to anchor in bodies, pianos, and sound.

There is now a long tradition of playing Pärt's music with various instrumentations; *Fratres*'s seventeen different versions published by Universal Edition is the most striking example. When asked about this particular per-

formance practice, Pärt frequently provides neo-Platonic explanations; he asserts that music's highest value is outside its color and exists beyond particular instruments.[26] In a sense, Pärt's position echoes the attitude, described
by Skowroneck, that understands instruments as subservient to a composer's
vision. Our own approach here resonates much more with the other camp
Skowroneck describes, which understands composition as arising from the
material possibilities of specific instruments. Through our analysis of cinematic representations of Pärt playing the piano and of one particular pianist's
process, we have grounded *Für Alina* and Pärt's music generally in the possibilities the piano offers. But what if *Für Alina* were to be played on the violin?

At the first "tintinnabuli" concert in Tallinn in 1976, *Für Alina* was performed without a piano. Two cellos maintain a drone on a low B for ten seconds before two violins enter playing the M- and T-voices of *Für Alina*, followed by the entrance of high-pitched chimes. The piece unfolds with drone
and bells continuing throughout until the violins reach measure 11, when the
cello drone and bells drop out and the violins play the last four measures on
their own.[27] On the one hand, this is recognizably the same piece; the same
melodies unfold, and the cello drone contributes to an overall increase in resonance for all the sounding instruments in the concert venue. Once the drone
and chimes drop out, the two violins by themselves do not echo as much, their
sound a bit drier. The live acoustic and at times imperfect playing stand in
stark contrast to the sound of recordings we have come to expect from Pärt and
Manfred Eicher at ECM. Yet even played with strings and chimes, with all
the quirks of live performance — and of a first performance of a new musical
idiom, no less — the importance of resonance is here from the beginning, the
cello and the resonating room serving an analogous function to the piano's
pedal.

If we hold on to the existential phenomenological claim that sensory experience creates meaning within a particular cultural and historical context,
the distance between *Für Alina*'s first performance and contemporary performance might seem vast. Isn't the nature of a string instrument's decay and
production of overtones created differently by the body? The score serves as a
reference point, providing a locus of convergence for composer and performers across history and geography. With Hillier's biographical information in
mind, we have considered how Pärt's relationship to the piano contributed to
his interest in wide ranges and resonance — musical features readily apparent
in *Für Alina*. The 1976 performance in Tallinn did not include piano, but the
addition of the cello drone, bells, and the ambient space creates an expression
of sustain, resonance, and decay.

So where does the "work" exist, if upon close examination different instruments, situations, and bodies create so many various pathways toward musical meaning? A piece of music cannot be a static thing but is iteratively produced and reproduced at each moment of performance (be it live or recorded), the somewhat unique result of particular times, places, performers, listeners, and instruments. In a sense, composer and performer meet in a virtual manner, both at turns occupying the piano bench. Different performers, in various times and places, inevitably experience learning a piece of music in relationship to their own particular habitus, context, and network of embodied memories. At the same time that certain elements of a work like *Für Alina* can remain consistent, Cizmic finds that Pärt's compositional strategies tap into her training in tonal piano repertoire but also that his emphasis on the instrument's resonating capacities seems to resist those habits and call for a different manner of playing.

Playing *Für Alina* (as experienced by Maria Cizmic)

Setting aside all of my habits and preconceived notions is impossible; I am a socially situated person, and my training as a pianist and musicologist (in addition to all my other experiences) mediates my perception of the new music I choose to learn. After bracketing simplicity, I notice other preexisting habits and ideas. First, I've listened to a lot of recordings of tintinnabuli music that influence what I think *Für Alina* should sound like. Perhaps more significantly, though, my experiences playing and listening to tonal music far outweigh my exposure to other musical idioms. And although I am well aware of how tintinnabuli draws on early music while subverting modernist practices, I found that my attention focused on how Pärt refigures aspects of tonality. Nora Pärt observes that playing Pärt's music is difficult because it is not typical of usual training and that musicians often need to produce sound in a new way.[28] This is most certainly true, and what I found particularly interesting was how Pärt's reformulation of tonal materials interacted with my conventionally classical training as a pianist. I found that compositional elements would call up habits from playing tonal music and then throw them open to question because in order to produce a different kind of sound, as Nora Pärt points out, I needed to refigure the interaction between my body and the piano.

The first thing I notice is my impulse to phrase (see Figure 1). For example, the initial bass notes plus subsequent melody imply i–V–i harmonic motion, which prompted a desire on my part to make the first four measures a phrase by creating a little crescendo that leads to the melody's high point on F#, then a decrescendo as the melody walks down to B, and then a longer pause in measure 4 to set this off from what comes next. But I'm quickly not sure if this is

Für Alina
für Klavier (1976)

Arvo Pärt
(* 1935)

Figure 1. *Für Alina*, mm. 1–5. Arvo Pärt "Für Alina | für Klavier" © Copyright 1990 by Universal Edition A.G., Wien/UE19823.

appropriate. There is no real harmonic motion, after all; the left hand never leaves the B minor triad, and in measure 4 the left hand ends on F#, undermining what might sound cadential.

I then consider how much force to use when playing the opening two-octave span on B. The louder these are played, the longer they last; but if these initial notes are too loud they seem to run counter to the ensuing spare texture. In either case, my perception of these notes returns me to Husserl's notion of retension.[29] In *Für Alina*, the whole piece occupies the aftermath of the initial B double octave, and I find it difficult to pinpoint when its sound has completely decayed, since its presence grounds the whole piece (at least until I lift the pedal in measure 11). The continued reverberation of the initial bass notes serves as a pedal point and could reinforce the lack of harmonic motion, or its lingering sound (or the impression that it lingers) might make the end of measure 4 sound a bit more like a cadence.

Other moments in the piece—most notably measures 8 and 11—also call up tonal habits by hinting at the dominant. In measure 8, the melody reaches its longest length and its highest register through a large leap upward and concludes with a minor third on F and A that implies the dominant. All of these elements prompt me to find some mini-drama here, to create a little delay in reaching that high D, slowing down to set apart the minor third on F. And here I find that the process works against some of the tonal implications it creates. Measure 11 (where Pärt drew a small flower in his original score) brings the arrival of C# and F, the most clear reference to the dominant.[30] But by this point *Für Alina* is already winding down—the melody is shorter and has retracted closer to its opening register—and these factors prompt me to resist making too much of C# and F as a dominant that resolves to the tonic of the last four measures.

These references to tonal practice, and my attempts to negotiate my own habits, are further attenuated by Pärt's indication to hold the pedal down from measures 1 to 11, coupled with the role of tempo and duration. It is the pedal that fundamentally shifts my approach to phrasing and articulation. For example, to make the first four measures a phrase, I try playing the whole statement legato; that choice, I discover, dampens the amount of overtone-rich resonance possible with the pedal continuously held down (which all the ECM recordings have taught me to value). If I could clear the pedal, I would be inclined to do so at the end of measure 4, in the middle of the F to high D leap in measure 8, and on the C# to F in measure 11, and in this way mark out what I hear as the mini-drama of *Für Alina*—its attempt to grow and recede, to go somewhere and return. Instead, holding the pedal down continuously prompts me to pay attention to the way each pair of notes resonates. I opt for a detached articulation and depress and release each pair of notes simultaneously and for less dynamic variation; this approach seems to increase the overall resonance of my piano. These choices, I believe, flatten out the tonal implications of the composition.

As I play, I spend a lot of time concentrating on making sure that the articulation is totally even in my right and left hands, on making the melody ring out, on avoiding wrong notes and then cringing when they inevitably occur. Moments when the melody is not voiced well and drops out and moments when wrong notes show up seem extremely obvious and disruptive. The options for correcting these mistakes are unsatisfactory—lifting the pedal clears out the wrong note and everything else with it, erasing the piece's ringing effect. Keeping the pedal down, possibly the better option, makes the wrong note lasts for a while until it becomes just another resonating overtone. The conductor and composer Andreas Peer Käler writes that his attempt to give *Für Alina* to a beginning piano class was an utter failure and that Pärt's seemingly simple score holds an "astonishing potential for peril."[31] Käler might be overstating his case, but the texture of this piece is incredibly fragile. Even if a wrong note does not impose itself, some ambient sound might prove similarly disruptive — there is always the sense that some intrusive sonic event could rattle *Für Alina*'s delicate balance.

It eventually occurs to me that though I am concentrating on many things (phrasing, articulation, reverberation, playing the right notes), I am not really focused on the compositional process. As Don Ihde argues, a person makes choices regarding which aspects of experience, and which sounds, will occupy one's attention.[32] Pärt tells Restagno that one reason for the popularity of *Cantus* is that one can clearly hear the structure on the music's surface.[33] Perhaps the habits of composers (not to mention of music scholars) emphasize structure, while the habits of performers and audiences are different. Ihde notes that there are many entrenched habits that mediate what we pay attention to and how we experience the world.[34] For me as I play, it is very easy not to focus on the "architecture" of the piece — as I concentrate on all these other elements, I somewhat peripherally notice a general swell in the length of melodies and their reverberations that then diminishes, particularly with the last four measures after the release of the pedal. Other elements stand out, such as the repetition of three notes in the left hand and large leaps in the right hand (especially the dramatic leap in the middle of the piece at measure 8). The transparency of the process leads me to find that the process is less important than its byproducts — texture, tempo, resonance, overtones, diatonic dissonance.

Pärt's handling of diatonic dissonance constitutes one of tintinnabuli music's most striking characteristics. Through a phenomenological lens, certain dissonant intervals stand out more than others depending on their location.[35] For example, the first major second in measure 2 is very obvious (see Figure 1, again); the same B–C♯ interval occurs in the middle of measure 6, but there the dissonance seems less prominent given its position in the middle of the melody. I may not be so aware of the process, but other elements that might be included in the concept of form — like the beginnings and endings of melodies and their relative consonance/dissonance — stand out in my perception. The width of the intervals, all displaced by an octave, seems crucial. The effect of a major second is much harsher than that of a ninth; the wider interval seems tense but not biting. Because the pedal is held continuously, I lose the more specific sense of a dissonant interval moving to a consonant one. The minor tenth in measure 2 contains the continuing reverberation of C♯, so the movement from relatively dissonant to consonant intervals does not have the same quality of tension and resolution I might expect based on my habits of tonal listening and playing (see Figure 1, again). And, as the melodies unfold, all the notes feed into the ringing of overtones, which contains many close intervals. A sonic continuity results between the pairs of notes as they arrive and the aura around them, generating a fairly constant space of tension.

The combination of slow tempo, quiet dynamic level, and continuous use of the damper pedal in *Für Alina* foregrounds the decay of sound. This musical feature returns me to Ihde's discussion of attention and what he describes as a focus/field/horizon model.[36] Everything one could possibly hear constitutes the auditory field; within that field one can choose what to focus on, and one's focus can change. There is a horizon, or a limit, to that field — for example, I can hear my colleagues chatting down the hall, but I cannot hear the students walking outside my building. Even though there is no absolute silence in this world, we might consider that the decay of sound and return to relative silence (or absence of a particular sound) constitutes a horizon for music, in a general sense. Typically, listeners focus on the sounding of music rather than the decay of that sound; with this in mind, Pärt takes the peripheral experience of sonic decay and encourages a listener to focus on it (making it no longer a limit experience). Ihde considers the phenomenon that music seems to come out of and then dissipate into silence, and he writes: "Through the creation of music humans can manipulate the mysteries of being and becoming, of actuality and potentiality, and through the vehicle of music they can legislate the schedule of a phenomenon's passage from its total being to its absolute annihilation."[37] Although Ihde's phenomenological exploration of listening is fruitful in many ways, here he moves away from embodied experience to metaphor. The bodily creation and cessation of music can be heard metaphorically, in this case as performing the existential frame of life and death, being and becoming. This assessment of music generally speaks to the "simplicity is profound" narrative that often encircles Pärt and his music. Simplicity in and of itself may not be profound, and not all music is about life and death, but paying attention to small combinations of notes as they sound and then fade away can serve as an aural metaphor for metaexistential issues regarding life and death; this may supply some grounding in the music as to why many people hear Pärt's music in this way.

To return to Heidegger's point about sound and context, in my experience working through *Für Alina* I found that my own classical piano training and experience with tonal music emerged as a primary context. Perhaps a musician steeped in early music would find a different set of bodily habits and musical knowledge activated. The expressive possibilities that arose — the push and pull of various sorts of tension, of fragility, and of sonic decay — came out of the physical negotiation between instrument and body, both as repositories of past music I have heard and played. As musicians and music scholars we are conventionally taught to understand tonality as a system of harmony and form; through the process of this phenomenological analysis we also conclude that tonality is constituted by the many habits of bodily training and performance. These habits may not apply in the case of tintinnabuli, and as a result Pärt's reformulation of a single key area creates tension — elements such as melodic motion trigger expectations that end up being attenuated at best, leading to a kind of ambivalence for the performer/listener.

Conclusion

By setting aside the concept of simplicity, we have sought to demonstrate the ways that past experience playing and listening to tonal music constitute a habitus and context through which *Für Alina*'s expressive possibilities arise. The discourse of simplicity that encircles Pärt's music gets used to make value judgments, and as such it does not describe the musical experience all that well. Instead, a phenomenological approach puts pressure on bodily musical experiences that give rise to expressive possibilities. Instead of simplicity, we found a sense of tension from thwarted expectations based in tonality and an overall texture that conveys fragility. The importance of the sustain pedal emphasizes resonance, which focuses attention on duration and blurs the distinction between tension and release. The use of diatonic dissonance coupled with reverberation from the damper pedal elides any sense of possible resolution to consonance and emphasizes tension while shifting focus to the decay of sound. To close the hermeneutic circle depends on understanding these bodily and expressive experiences in their various possible historical and cultural contexts; with this in mind we can point in a few directions. Given the late-socialist, East European context of tintinnabuli's birth, how might the expressive possibilities identified here be understood in terms of the search for paths out of both socialist realism and modernism? How might we understand the relationship between the resonant spaces of unresolving tension in relation to the other streams that fed into the development of tintinnabuli, such as early music and religion? And do these embodied expressions of fragility and the focus on the decay of sound figure into the appeal Pärt's music held for the younger generation of Soviet underground composers in the 1970s? By the mid- to late 1980s, Pärt had put his own experiences as a recording engineer to use in his collaborations with Eicher at ECM, and a parallel set of questions could be followed through regarding the Western reception of tintinnabuli music. In particular, we might consider how the musical expression of fragility and the focus on the decay of sound provide a strong musical explanation for the practice of listening to Pärt's music in contexts related to illness, death, and dying.[38] Here we only gesture toward ways that an analysis of embodiment can relate to historiographical and hermeneutic questions regarding Pärt's music. In this way, a phenomenological approach can help point to the plurality of coexistent meanings that dance around tintinnabuli.

Notes

1. Harris Berger, "Phenomenological Approaches in the History of Ethnomusicology," *Oxford Handbooks Online*, Oxford University Press, http://www.oxfordhand books.com.

2. Maurice Merleau-Ponty, *The Phenomenology of Perception* (London: Routledge, 1962), 103–7.

3. Matthew Rahaim, *Musicking Bodies: Gesture and Voice in Hindustani Music* (Middletown, CT: Wesleyan University Press, 2002), 17.

4. Elaine King and Anthony Gritten, *Music and Gesture* (New York: Routledge, 2017), xxi.

5. David Sudnow, *Talk's Body: A Meditation between Two Keyboards* (New York: Knopf), 6–7.

6. Enzo Restagno et al., eds., *Arvo Pärt in Conversation*, trans. Robert Crow (Champaign, IL: Dalkey Archive, 2012), 29.

7. Restagno et al., *Arvo Pärt*, 37.

8. Restagno et al., *Arvo Pärt*, 37.

9. Jeffers Engelhardt, "Perspectives on Arvo Pärt after 1980," in *The Cambridge Companion to Arvo Pärt*, ed. Andrew Shenton (Cambridge: Cambridge University Press, 2012), 34–35.

10. Wolfgang Sandner, CD liner notes to Arvo Pärt, *Tabula rasa* (ECM 1275, 1984), 19.

11. Laura Dolp, "Arvo Pärt and the Marketplace," in *The Cambridge Companion to Arvo Pärt*, ed. Andrew Shenton (Cambridge: Cambridge University Press, 2012), 189.

12. Robert Sholl, "Arvo Pärt and Spirituality," in *The Cambridge Companion to Arvo Pärt*, ed. Andrew Shenton (Cambridge: Cambridge University Press, 2012), 149–50.

13. Geoff Smith, "An Interview with Arvo Pärt: Sources of Invention," *Musical Times* 140, no. 1868 (Autumn 1999): 21.

14. Restagno et al., *Arvo Pärt*, 60.

15. Peter Schmelz, *Such Freedom, If Only Musical: Unofficial Soviet Music during the Thaw* (New York: Oxford University Press, 2009), 6–13, 216, 220–21.

16. Don Ihde, *Listening and Voice: Phenomenologies of Sound* (Albany: State University of New York Press, 2007), 219.

17. Tilman Skowronek, *Beethoven the Pianist* (Cambridge: Cambridge University Press, 2010), 162, 167.

18. Emily I. Dolan, *The Orchestral Revolution: Haydn and the Technologies of Timbre* (Cambridge: Cambridge University Press, 2013), 8–10.

19. Martin Heidegger, *Being and Time*, trans. John Macquarrie and Edward Robinson (Malden, MA: Blackwell, 1962), 278.

20. Restagno et al., *Arvo Pärt*, 26–27.

21. Paul Hillier, *Arvo Pärt* (Oxford: Oxford University Press, 1997), 26.

22. See Engelhardt's chapter in this volume.

23. Heidi Maria Lehtonen et al., "Analysis and Modeling of Piano Sustain-Pedal Effects," *Journal of the Acoustical Society of America* 122 (2007): 1781–97.

24. Herman Sabbe, "Music Makes Time — Music Takes Time: Apropos of Arvo Pärt's *Tabula Rasa*," *New Sound* 17 (2001): 47–51; Sholl, "Arvo Pärt and Spirituality," 156.

25. Restagno et al., *Arvo Pärt*, 58.

26. Martin Elste, "An Interview with Arvo Pärt," *Fanfare* 11, no. 4 (1987–1988): 339.

27. Thank you to Kevin C. Karnes for providing a recording of this performance. See also Kevin C. Karnes, *Arvo Pärt's* Tabula Rasa (New York: Oxford University Press, 2017), 48–49, for a reproduction of this concert's program.

28. Restagno et al., *Arvo Pärt*, 78–79.

29. Ihde, *Listening and Voice* 89; see also Sabbe, "Music Makes Time," 49–50, in which he also discusses retention.

30. Hillier, *Arvo Pärt*, 88–89, includes a reproduction of the autograph score of *Für Alina*, with flower.

31. Andreas Peer Käler, "Radiating from Silence: The Works of Arvo Pärt Seen through a Musician's Eyes," in *The Cambridge Companion to Arvo Pärt*, ed. Andrew Shenton (Cambridge: Cambridge University Press, 2012), 194.

32. Ihde, *Listening and Voice*, 18, 74–75.

33. Restagno et al., *Arvo Pärt*, 39.

34. Ihde, *Listening and Voice*, 74.

35. My observations here resonate with Andrew Shenton, *Arvo Pärt's Resonant Text: Choral and Organ Music, 1956–2015* (Cambridge: Cambridge University Press, 2018), 43–46, where he describes using the open source Music21 software platform to analyze the degrees of dissonance and tension both vertically and between neighboring notes in Pärt's organ music.

36. Ihde, *Listening and Voice*, 73–77.

37. Ihde, *Listening and Voice*, 223.

38. Kythe Heller, "An Ethnography of Spirituality," in *Arvo Pärt's White Light: Media, Culture, Politics*, ed. Laura Dolp (Cambridge: Cambridge University Press, 2017), 122–53.

V

Theology

11
Presence, Absence, and the Ambiguities of Ambience
Theological Discourse and the Move to Sound in Pärt Studies
Robert Saler

My goals in this essay are relatively modest and focused. Because I speak as a systematic theologian—and because I come to this task as a long-time consumer and devotee of Arvo Pärt's music rather than as a scholar of it—the most useful contribution I can make is programmatic and schematic. One of this volume's key questions is: What happens when "Pärt studies" moves intentionally into the field of "sound studies"? I seek to advance that question by addressing it through the lens of theology, specifically by suggesting debates in academic theology that might bear on this move to sound. I will also ask how the God-talk associated with Pärt-and-sound moves in the space between silence and resonance that Karmen MacKendrick locates as the home for theological discourse in her excellent book *The Matter of Voice: Sensual Soundings*:

> The willingness to dwell in those strange discursive places where theology happens requires a willingness to hear differently, to accept that language is doing something (rather than nothing) even as it struggles with its fall into silence. Into silence, and also into sound: wondering about saying requires attending to those who say, including the very strange (such as gods who somehow make by speaking) and the abjectly marginalized (such as criminals nailed up to die by suffocation); to the water in water movements of other animals; to the world itself, which might sign or sing or cry out; to citation as it edges into recitation as that in turn edges into liturgy, with its deliberately contagious quoting. If there is faith in theology, it is faith that we are not speaking mere nonsense, faith that there is something here worth listening to.

That something is a very abstract something most of the time. But it is
a very concrete something, too. It resounds.[1]

The word "resonance" has implications for our method. In pedagogy within
the humanities, for instance, Adam Davis has argued that intertextual and interdisciplinary work sometimes proceeds best on the basis of leaving aside "relevance" in favor of resonance.[2] For Davis, to focus on the "relevance" of a text
of a given conversation tends to direct people into their normal disciplinary
talking points. The same can be said, for our purposes, about a musicological
discussion. The art is to find texts that are "resonant" with a situation. That is to
say, they are just enough off to the side that they disrupt the normal strategies,
tactics, and talking points that constrain imagination yet remain just enough
connected to the topic that they can provoke and evoke previously unseen
points of contact. So, the claims in this paper have less to do with the modes
of discourse to which Pärt-and-sound might be relevant and more to do with
how the shift from silence to sound in Pärt studies creates new possibilities for
resonance with theology.

Theology tends to be discursive. The sounds it makes are speech (or, if written, the words are usually meant to be read as prose). Music is neither discursive nor spoken prose. But if the questions from one mode of sound (speech) to
another (music) can resonate in this technical sense, then we are on relatively
firm scholarly ground in an inquiry between theology and music.

That shift away from relevance to resonance agrees with emerging interdisciplinary methodologies within sound studies itself. As Jonathan Sterne argues,

> Sonic imaginations denote a quality of mind, but not a totality of
> mind. In addition to carving out their own intellectual spaces within
> other fields, sound students facilitate the sonic imaginations of schol-
> ars who might deal in sound in their work even though it is not their
> primary concern. Just as concepts of the gaze and images bounce back
> and forth between studies of visual culture and much broader fields of
> social and cultural thought, so too do concepts with a sonic dimension
> like hearing, listening, voice, space, and transduction (to name just a
> few) — and sound itself.[3]

Sound studies as a discipline is already an interdiscipline that has among its
goals a kind of catalyzing of critical attention. It aims to show how sound
functions both as substance and cipher within discourse. It is an interdiscipline of strange bedfellows by design, yet thus far it has not especially wel-
comed theology among them. In his recent book *Against Ambience*, Seth Kim-
Cohen follows the lead of much critical theory in portraying theology as the

disembodied abundance of unearned assurance of presence. Such assurance inevitably betrays its own uncertain foundations by discursively (or literally during theocracies) controlling access to valid transcendence.[4] This theology is not the space that Karmen MacKendrick proposes, in which meaning meets elusiveness; rather, it is power mystifying itself as revelation. There is no doubt that a great deal of theological method, analysis, and reflection — especially through the regimes of sword, dogma, or rationalism — is of little use to critical sound studies. Even less would most theological methods, analyses, and reflections be useful for drawing out insights from a composer such as Pärt, however explicitly religious that composer may be or how sacred his works. But a cynical, theological parody of presence is only one option, and, indeed, in the tradition it has always been just one option. Here, I will outline a more salutary alternative.

Falls into Sound

Musicological and theological studies of Pärt to date have proceeded largely along the lines laid out by MacKendrick's identified dynamic of speech that gives way to silence. In Peter Bouteneff's masterful work *Arvo Pärt: Out of Silence*, the theologically rich interplay between silence and sound provides a vantage point from which to view the iconic character of Pärt's work. Bouteneff writes about how it relates to Christian reflection historically: "The purpose of silence is to generate a right word, a right sound. There is not one without the other: no silence without sound, and no sound that does not begin or end with, or somehow embody, silence."[5]

I suggest that studying Pärt in light of the sound focus creates new avenues of engagement with Christian theology's long-standing ambiguities regarding the question of presence and absence — and the constructive interplay between those two. By "new avenues" I do not imply a radical departure from the foundation already laid by the work of Bouteneff and others. Instead, I think we are seeing the possibilities for a particular kind of focusing of the questions. And the questions would simply be these, to start: Are Pärt's music and its resonance with theology a constructive contribution to our thinking about divine presence and absence? Conversely, are the categories of presence and absence helpful theological hermeneutical tools to elucidate something interesting about not only Pärt's music but also the Pärt phenomenon as it plays out in aesthetics, culture, and consumption?

To begin to construct affirmative answers to these questions, we should first note that the presence/absence dialectic in theology, at least as it pertains to God, is about more than static conceptions of presence as consolation, spiritual experience, or insight. It is also about more than their converse: spiritual aridness, evil, or forsakenness. To stay in a simplified binary—whereby God's presence is somehow unqualifiedly good and divine absence is simply lamentable (however necessary or inevitable)—is to gloss over important nuances within the Christian tradition past and present.[6] The work of phenomenologists, such as Jean-Luc Marion, who study the intersection of the idol and the icon reminds us that the revelation of divine presence conceals as much as it reveals. (And it does so largely on the basis of excess that can in fact be disruptive and even terrifying.) The distinction here is between the idol (that which arrests the gaze in a false mastery of the divine) and the icon (that which conveys the gaze to the God that exceeds every image and sound). "Access to the divine phenomenality is not forbidden to man; in contrast, it is precisely when he becomes entirely open to it that man finds himself forbidden from it—frozen, submerged, he is by himself forbidden from advancing and likewise from resting. In the mode of interdiction, terror attests to the insistent and unbearable excess in the intuition of God."[7]

In the spirit of Exodus 33–34, where Moses encounters God on Sinai but is allowed to see only his back (otherwise he would have been destroyed by divine presence), we might envision God's absence or veiling precisely as a mode of divine benevolence. This benevolent absence may aim to protect us from destruction—a concern that is not allayed simply by our seeking to overcome ontotheology or metaphysics as such. The absence also guards against the sort of delusions of comprehension that cannot but lead to conceptual idolatry and attendant violence. A divine desert is more fecund than a marketplace of idols.

We must retain a tense and shifting relationship between the twin dialectics of presence/absence, healing/destruction (or deconstruction), life/death, etc. Moreover, we must not align the two binaries too strictly if we are to do justice either to Christian theology or to conversations around how we might locate the work of a given composer within that matrix. We take note here, as have many before us, of the use of Pärt's music in hospice situations, medical centers, and in other settings where it is cited for its potential for healing, consolation, or the evocation of God's presence. And certainly, in popular discourse it is understandable that much talk of the beauty of Pärt's music depicts it as a kind of access to an ethereal mode of divinity whereby God's beauty and being are somehow made available to listeners in a relatively straightforward aural connection between sound and divine presence.[8] That's not wrong, as far as it goes, and in fact it can be theorized quite productively.

The Irish Catholic philosopher William Desmond can be helpful here. At the heart of his work is an investigation into the relationship between the particular and the universal, the part and the whole, the immanent and the transcendent. To Desmond, the task is to navigate this relationship: Rather than understanding particularity as an instance of discrete identity, Desmond argues that it emerges from an original overabundance or surplus of being. That is, there is no ontological sense to the notion of isolation. All existence is imbued with a constitutive porosity, an internal relatedness, to the universal. Nothing is truly isolated. Desmond claims that an openness to the universal lies at the ground of, and is intimate to, the particular.

Indeed, Desmond contends that religion opens us to the question of the universal and our particular relation to it. Further, the universal "communicates" intimately with the particular through the aesthetic. Desmond explains, "The often sterile opposition of universal and particular must be questioned. . . . We cannot deny something that touches us intimately in art, and yet the intimacy is not closed into isolated singularity or subjectivity. Quite the opposite — something is communicated in the aesthetic intimacy that radiates beyond such closure."[9] The aesthetic communicates the universal in/at the very depth of the particular.

As one might imagine, for Desmond, music is a particularly privileged mode of aesthetic porosity:

> Music as form that is forming and formless, as intimate and yet more than itself, as universal speaking to all, even those who resist. It reminds one of the graced porosity of true prayer — one is taken by the music. It is not a matter of rational self-determination. . . . Song is self-communication, yes, but in the selving there is more than self communicated. And hence in the *passio* of the music there is more than I or you, for again it is the music that envelops us, comes from below up, from above down, comes from no particular side and from every side. One is invaded and one is caressed. One is carried away and yet, as in the surrender of love, one is peacefully at home with oneself and the flow.[10]

As we have seen elsewhere in this volume, the move to sound in Pärt studies can encompass the materiality of sound and the intimacy of presence between aesthetics, religion, and experience in fruitful ways and along the lines that Desmond suggests. I would want to argue that — alongside these strategies — the most interesting intersection between theology and sound in Pärt might come if we look at that intersection as an embrace of the same ambiguities of divine presence characteristic in theology itself — apophasis and kataphasis — the unsaid and the said[11] — in mutually reinforcing motion across the physicality of embodied hearing. And this would resonate precisely to the extent that Pärt's oeuvre presents a particularly compelling instance of music that in its very sound embodies presence — in precisely this concealing, retiring, hidden sense. We might even stipulate that, in music as in theology, it is only when presence is also absence that presence is something other than delusion, or idol. In other words: If Pärt's music mediates the divine in such a way that presence is complicated rather than assumed, then Pärt's music is even more theologically interesting than we thought; it is recognizably a sonic embodiment of MacEndrick's previously cited "pulls of tension that give us delight." This insistence that the sonic presence-ing of absence serves as a partial but important safeguard against idolatry will set the theological agenda for what is to follow in this chapter.

Stubborn Material

To fill out the picture of what this dialectic looks like theologically, we can move our inquiry to a space that seems far removed from our topic — Nazi Germany.[12] From 1935 to 1937, the German Lutheran theologian Dietrich Bonhoeffer — having been stripped of his license to teach theology at the University of Berlin because of his resistance to the Nazification of the theology faculty there, as well as the Nazification of the broader Protestant and Roman Catholic church in Germany — served as founder and principal of an illegal seminary at Finkenwalde serving the Confessing Church. During his two years there, the ecumenically inclined Bonhoeffer began modeling the life of the community along the lines of monasticism: hours of daily prayer, readings at meals, psalm chanting, and so on. One of his most striking additions to the rhythm of the community, however, was the reintroduction of individual confession and forgiveness, as opposed to the more corporate model, which by that time was firmly ensconced in Protestant practice.

As he recounts his own theology behind the practice in *Life Together*, his memoir of the Finkenwalde years, Bonhoeffer offers an astute psychological and theological portrait of how embodiment and materiality keeps divine presence from slipping into idolatry. He writes:

> In confession there occurs a breakthrough to assurance. Why is it often easier for us to acknowledge our sins before God than before another believer? God is holy and without sin, a just judge of evil, and an enemy of all disobedience. But another Christian is sinful, as we are, knowing from personal experience the night of secret sin. Should we not find it easier to go to one another than to the holy God? But if that is not the case, we must ask ourselves whether we have often not been deluding ourselves about our confession of sin to God — whether we

have not instead been confessing our sins to ourselves and also forgiving ourselves. And is not the reason for our innumerable relapses and for the feebleness of our Christian obedience to be found precisely in the fact that we are living from self-forgiveness and not from the real forgiveness of our sins? Self-forgiveness can never lead to the break with sin. This can only be accomplished by God's own judging and pardoning Word.

He goes on:

Who can give us the assurance that we are not dealing with ourselves but with the living God in the confession and forgiveness of our sins? God gives us this assurance through one another. The other believer breaks the circle of self-deception. Those who confess their sins in the presence of another Christian know that they are no longer alone with themselves; they experience the presence of God in the reality of the other. . . . The other Christian has been given to me so that I may be assured here and now of the reality of God in judgment and grace. As the acknowledgment of my sin to another believer frees me from the grip of self-deception, so, too, the promise of forgiveness becomes fully certain to me only when it is spoken by another believer as God's command and in God's name.[13]

Note the importance of the speech-event and hearing in this encounter. The auditory character of Bonhoeffer's argument here is no accident. While the history of Protestantism subsequent to the so-called magisterial Reformation often is linked with the importance of the written text of the Bible, at least for the Lutheran tradition in which Bonhoeffer was formed the greater emphasis was actually upon the spoken word and the divine presence within it. According to Luther (and later picked up by such Reformed theologians as Karl Barth), whether in preaching or in the words of forgiveness, Christ is in fact ontologically present in benevolent and salvific fashion. Indeed, the Augsburg Confession (Article 7) defines the church as an event of audition: The church is present wherever the word is rightly preached and the sacraments rightly administered. For Bonhoeffer, the Catholic practice of private confession and absolution is, in some ways, a return to the Reformation's emphasis on *grace as embodied sound*.

This may also seem like a return to naïve "presence-ing," but adding a few layers of complication from Bonhoeffer's own context can address that concern. While Bonhoeffer, like his mentor Barth, placed the church at the center of his theology, biographically we should notice that he was ecclesially homeless for virtually all of his life. He was caught first in the throes of Harnackian liberal Protestantism, which left church and theology largely as ornaments and brokers of bourgeois citizenship ethics in which the state supplied the content of Christian discipleship. Then he was caught in the wake of the fascism that he considered the inevitable consequence of anemic theological and ecclesial practice. Bonhoeffer's entire case against the silence of confession as solo prayer rests on the accusation of projection. That is, if confession only takes place in silent prayer (or its corollary, the anonymous audition of rote corporate confession and forgiveness in the liturgy), then the God to whom we pray is plastic to our projections. This analysis coheres with the silence to which the sound theorist David Toop (like Kim-Cohen, no fan of theology) attributes to too much Christian practice:

> There are silences of peace, and then there are silences of complacency, stasis, regulation, piety, submissiveness, secrecy, ostracism, excommunication, the status quo, a deserted town center after dark, gloomy Sunday, a gated community, suburbia, a cold church pew, people living quiet, respectable lives or suffocating under ennui, shame, embarrassment, inhibition, boredom.[14]

Or, we might add, the silence of the reduction of prayer to projection. This is Bonhoeffer's great insight: Silence as well as sound can mistake the idol for the icon.

God gives the confessor as a gift to the confessing one precisely because the "other" is not plastic to our projections. The materiality of sound in the confession and (derivatively) authoritative confidence in proclamation of forgiveness is stubborn. For Bonhoeffer, the move to sound is the move away from projection, even though that move can never happen perfectly or completely. That move mediates between the absence of God and the presence of salvation through the sound of absolution.[15] Sound here encompasses presence and absence in a salvific, fleshly cadence. The rupture of projection for the sake of the external word justifies. Sound is needed, but it must be sound that encompasses the God who refuses to be fully present (since such full presence can only be that of the idol, not the icon).

Sounding the Absence

So what then of Pärt? If I am correct in arguing that theological discourse in its nonidolatrous mode bears within its speech both concealment and revelation, absence and presence, then we might have a way to envision how the move to sound in Pärt studies extends some fruitful lines of inquiry into Pärt and theology that have been inaugurated by Bouteneff and others. We can take, for instance, the question of the interplay of sound and silence in Pärt as it pertains to the oft-cited critique of scholars like Jeremy Begbie, who in some sense mirrors Kim-Cohen's disdain for depoliticized ambience: "The music of composers such as Taverner and Pärt . . . is characterized by a highly contemplative ambience and is often labeled spiritual . . . it offers a cool sonic cathedral in a hot, rushed, and overcrowded culture." But Begbie goes on to argue that such sonic cathedrals are insufficiently embodied, insufficiently incarnational. He asks:

> If Christ has embraced our fallen humanity, including its fear, anxiety, hunger, loss, frustration, and disappointment, and these have been drawn into, indeed, become the very material of salvation, can we be content with a vision of the spiritual that is unable to engage just these realities, with a cool cathedral that bears little relation to life on the streets?

Begbie goes on to spell out (in terms that he may have intended to invoke Bonhoeffer) that such a vision of the spiritual is a "world-denying distortion of Christianity."[16]

To be sure, Bouteneff, Robert Sholl,[17] and others have usefully complicated this narrative, as do several chapters in the present volume. One can indeed question not only the musicology but indeed the Christology that Begbie employs in his diagnosis of Pärt. I would want to follow Shenton's invocation of "metamodernism" in relation to Pärt, arguing that "what Pärt has done, both with *Passio* and with 'tintinnabulation' in general, is to reconstruct music from the myth of the golden age of music, situating it in a metaxis that is neither ironic pastiche nor avant-garde."[18] I want to argue that the presence/absence dialectic I have outlined here can also gloss this same metaxis, this state of in-betweenness, as the refusal both of uncomplicated presence (idol) and of sheer apophatic silence. Far from building a cool escapist haven, Pärt creates music that presences absence. I take his way of conceptualizing his technique theologically to be one of the most fruitful intersections of sound and post/meta/modern theology. Better: It is theology as it has always been understood by theologians captivated by the refusal to domesticate transcendence.[19]

Pärt takes the risk of a sound that has space, an ambience that is incarnational precisely in its refusal to uncomplicate divine presence. I think that this is a promising direction for the theological study of Pärt as it moves into sound. The key is to ask the question of how Pärt's compositional technique and how the experience of it sounding upon ears — religious and unreligious alike — create space not only for us to insert our own narratives of meaning (as both Sholl and Shenton suggest) but for the sound itself to resonate with theology at its best. Such would be a theology that itself lives in this metamodern metaxis between sound that saves by sounding and sound that saves by withdrawing. We can agree with Desmond and countless Pärt fans that his music presents a "graced porosity" between God and our ears. But the grace comes at least partly from the fact that tintinnabulation, essentialism, and simplicity of sound take away as much as they give — just as revelation itself, and any words about it, must also do.

Notes

1. Karmen MacKendrick, *The Matter of Voice: Sensual Soundings* (New York: Fordham, 2016), 127–28.

2. See Adam Davis and Elizabeth Lynn, eds., *The Civically Engaged Reader* (New York: Great Books Foundation, 2006). Adam Davis is the former director of the Project on Civic Reflection, which brings nonprofit employees, from executive directors and CFOs to custodians and receptionists, together around key poems and literary texts in order to foster conversations around mission and vocational meaning.

3. Jonathan Sterne, "Sonic Imaginations" in *The Sound Studies Reader*, ed. Jonathan Sterne (London: Routledge, 2012), 9.

4. Seth Kim-Cohen, *Against Ambience and Other Essays* (New York: Bloomsbury, 2016), esp. 20–23.

5. Peter C. Bouteneff, *Arvo Pärt: Out of Silence* (Yonkers, NY: SVS, 2015), 130.

6. See Bouteneff, *Out of Silence* 108ff., for a helpful discussion on the motifs of "God-forsakenness" in two of Pärt's spiritual influences, St. Silouan the Athonite and Elder Sophrony.

7. Jean-Luc Marion, "In the Name," in *God, the Gift, and Postmodernism*, ed. John D. Caputo and Michael J. Scanlon (Bloomington: Indiana University Press, 1999), 41.

8. See also Albert Blackwell's work on the "intimacy" of transcendence in music: "Whereas vision entails interactions between our eyes and outer objects on which we focus, our organs of hearing, the middle and inner ear, are palpably more inward, their sensitivity is more omnidirectional, and their selectivity is less delineating. Thus our hearing conveys intimacy and immersion. Indeed, the sense of intimacy is even nearer. Our inner ear perceives sound not only by means of the outer ear but also from vibrations within our skull, and we feel sound, especially low frequency vibrations such as those of drums, bass organs, and large organ pipes, in our chest and abdomen. As mystical religious experience conveys immediacy and immersion in cosmic rhythm, dissonance, and harmony, music conveys immediacy and intimacy

in their sonic equivalents." Albert Blackwell, *The Sacred in Music* (Louisville, KY: WJK, 1999), 215.

9. William Desmond, *The Intimate Universal: The Hidden Porosity among Religion, Art, Philosophy, and Politics* (New York: Columbia University Press, 2016), 8. My thanks to Nick Buck for instructing me on the complexities of Desmond's project.

10. Desmond, *The Intimate Universal*, 91–92

11. "The God of Judeo-Christian tradition is inconceivable and indescribable, and must be approached in silence and darkness — the early Christian writers called this *apophasis*. Yet it has been given to humans to say something, to him and about him, which they called *kataphasis*." Bouteneff, *Arvo Pärt*, 129.

12. Given the omnipresent possibilities of fascism in Pärt's biography, and perhaps our own, it may not be so far away.

13. Dietrich Bonhoeffer, *Life Together*, trans. Daniel W. Bloesch and James H. Burtness (Minneapolis, MN: Fortress, 1996), 112–13.

14. David Toop, *Sinister Resonance: The Mediumship of the Listener* (New York: Continuum, 2011), 219.

15. This precise mediation between absence and salvation would later be theorized in Bonhoeffer's "religionless Christianity," the stubborn material worldliness that forsakes any God other than the one made secular, worldly, incarnate in Christ.

16. Jeremy Begbie, *Resounding Truth: Christian Wisdom in the World of Music* (Grand Rapids, MI: Baker Academic, 2007), 179.

17. See Robert Sholl, "Arvo Pärt and Spirituality," in *The Cambridge Companion to Arvo Pärt*, ed. Andrew Shenton (Cambridge: Cambridge University Press, 2012).

18. Andrew Shenton, "For Whom the Bells Toll: Arvo Pärt's *Passio*, Metamodernism, and the Appealing Promise of Tintinnabulation," in *Contemporary Music and Spirituality*, ed. Robert Sholl and Sander van Mass (New York: Routledge, 2017), 33.

19. For a beautiful recent meditation on this absence/presence dynamic as it relates to imagery, see Natalie Carnes, *Image and Presence: A Christological Reflection on Iconoclasm and Iconophilia* (Palo Alto, CA: Stanford University Press, 2017).

12

The Materiality of Sound and the Theology of the Incarnation in the Music of Arvo Pärt

Ivan Moody

In this chapter I shall examine some of the ways that Arvo Pärt transmits, by means of sound, the idea of the Incarnation of Christ. Theology, it might be thought, must of necessity be transmitted in music simply by means of text. Pärt's understanding of the way music works, however, brings text into the realm of music in unexpected ways, so strongly aware is he of the theological potential of both sound and silence. Indeed, it is in his virtuosic handling of the fragile boundary between these two conditions that he manages to convey the theology of the Christian faith in a way that is simultaneously profoundly silent and joyously resonant.

In principio

"Sound is my word. I am convinced that sound should also speak of what the Word determines. The Word, which was in the beginning."[1]

These words by Pärt encapsulate in many ways the essence of his work. Sound is his word, and that sound also proclaims the Word, the pre-eternal God. This happens in many subtle ways, his work operating in the triangular space created by music, text, and, importantly, silence, a phenomenon that one might describe as perichoretic (in other words, like the coinherent relationship of the Persons of the Trinity). The composer has spoken of this space often, perhaps most famously in his often-quoted observation on his tintinnabuli technique, in which he says, "I have discovered that it is enough when a single note is beautifully played. This one note, or a silent beat, or a moment of silence, comforts me."[2] Comforting or not, silence is not only a point of origin for his music[3] but, paradoxically, also a space within which the music may resound. A relevant theological parallel might be made with the notion of the womb of the Virgin as being "more spacious than the heavens" (Πλατυτέρα των Ουρανών), in that within the confines of her human body she contains the Creator of the Universe. The iconic representation of the Virgin *Platytera* is also known as the "Virgin of the Sign," in reference to Isaiah 7:14, "Therefore the Lord himself shall give you a sign; Behold, a virgin shall conceive, and bear a son, and shall call his name Emmanuel."

In the image of the *Platytera*, then, we see an immediate visual representation of something that Pärt achieves aurally: the containing of infinity within a space defined by the human condition, or, in the case of the composer, the containing of the resonance of the melodic and harmonic representation of infinity within the absence of sound. But, in that music functions in time, such a representation cannot be static. The consequences of the Incarnation demand to be worked through so that the vastness of this event that transcends the boundaries of Heaven and Earth may be seen and explained: what St. Maximos the Confessor describes as "the mysterious self-abasement of the only-begotten Son with a view to the deification of nature, a self-abasement in which he holds enclosed the limits of all history."[4] Romano Guardini, in his masterly study *The Lord*, also considers how the Incarnation tears the fabric of silence:

> The Christmas liturgy includes these beautiful verses from the eighteenth chapter of the Book of Wisdom: "For while all things were in quiet silence and the night was in the midst of her course, Thy almighty word leapt down from Heaven from Thy royal throne . . ." The passage, brimming with the mystery of the Incarnation, is wonderfully expressive of the infinite stillness that hovered over Christ's birth. For the greatest things are accomplished in silence — not in the clamor and display of superficial inventiveness, but in the deep clarity of inner vision; in the almost imperceptible start of decision, in quiet overcoming and hidden sacrifice. Spiritual conception happens when the heart is quickened by love, and the free will stirs to action. The silent forces are the strong forces.[5]

If such words wonderfully describe the Incarnation, they also describe to perfection the creative act, born, as Guardini says, of the "deep clarity of inner vision." One might see this reflected in the structuring of the work by Pärt that deals most specifically with the Incarnation, *In principio*, entirely derived from the narrative at the beginning of the Gospel of St. John. While the first, second, and final movements, dealing with the eternal nature of the Word, of the "man sent from God," and of the Incarnation itself (*In principio erat*

Figure 1. Thirteenth-century icon of the Great Panagia *Platytera* (Our Lady of the Sign), from the Spassky Monastery in Yaroslavl, Russia. Tretyakov Gallery, Moscow. Public domain image from Wikimedia Commons https://commons.wikimedia.org/wiki/File:Oranta.jpg.

Verbum, Fuit homo missus a Deo, and *Verbum caro factum est*) are robustly static, provoking Andrew Clements, in his rather negative review of the work,[6] to compare them to Philip Glass, the third, *Erat lux vera*, contemplating the nature of the True Light, and the structural center of the work, as well as being its longest movement, takes us to an area of more familiar Pärtian intimacy and quietude, suggesting precisely that the strength of that light, the *phos hilaron*, arises not through an ostentatious manifestation of power but through the assumption of the frailty of human nature — in other words, the center of the paradox of the Incarnation.

For Dale Nelson, in a wide-ranging discussion of music by Pärt and writings about him from 2002, the Incarnation, in Pärt's music, is something that may be intuited, but it is not put there consciously:

> For avowedly Christian music it is not, perhaps, very plainly founded upon the Incarnation of the Son of God. Yet the Incarnation is suggested in Pärt's work — not only by the intrinsic content of some texts, but also by the music's "awareness" of suffering.
>
> But the Incarnation is only suggested. For the witnesses of the Incarnate Savior, for the apostles, the opposite of worldly chatter, of Caesar's corruption of language, is not so much silence, but right speech — the prayer of the silent publican standing afar off, but also praise, and proclamation to the world, as the believer's response to the God who has already come to us in the flesh, and who continues to come to us.[7]

There is truth in these observations, but the Incarnation does not need to be shown in an obvious, pictorial way in order to be present, if one understands it as a fulcral point of the history of Salvation, an event inseparable from, and an essential part of, the whole narrative of the life, death, and resurrection of Christ.

Crucifixus etiam

The Incarnation of Christ leads, in other words, inexorably to His Passion and Resurrection. Or rather, as Fr. John Behr has noted, it is through the lens of the Passion and the Resurrection that we come to understand the Incarnation:

> It is a stubborn fact, or at least is presented this way in the Gospels of Matthew, Mark and Luke, that the one born of Mary was not known by the disciples to be the Son of God until after the Passion, his crucifixion and resurrection. . . . Thus, to speak of the "Incarnation," to say that the one born of the Virgin is the Son of God, is an interpretation

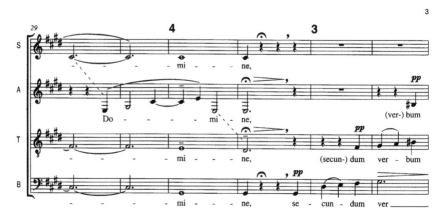

Figure 2. *Nunc dimittis*, mm. 29–34. Arvo Pärt "Nunc dimittis | für gemischten Chor a cappella"
© Copyright 2001 by Universal Edition A.G., Wien/UE31909.

made only in the light of the Passion. It is a confession about the crucified and exalted Lord, whose birth is then described in terms drawn
from the account of his death (the correspondence between the tomb
and the womb that delighted early Christians and is celebrated in liturgical texts and iconography).[8]

Such an approach has inevitable consequences for the artist, as Behr notes.
The artist, aware as he is of what will come after the birth of Christ, cannot
exclude that understanding from his depiction of that event. If Christ the Word
brings life, as St. Symeon recognizes, that life is nevertheless bought at the
price of His death. St. Ephrem the Syrian expressed it in the following words:
"By your resurrection you convinced them about your birth, for the den was
sealed and the grave was secured — the pure one in the den and the living one
in the grave. Your witnesses were the sealed den and the grave."[9]

     This awareness is clearly reflected in Pärt's treatment of the Song of Symeon,
or *Nunc dimittis* (2001), in which the word *verbum* ("word") is set to a single
bar, given to the tenors, utterly unlike the extensions and stretchings applied
to other words of the text. And the single bar introduces, after thirty bars of
hovering around a modal C-sharp minor with a dominant cadence, a B-sharp,
cadencing once more back in C-sharp minor, but this time subtly transformed,
tending toward, though not quite arriving at, F sharp (Figure 2). *Verbum*, in
other words, prepares the course of the rest of the piece, which moves toward
an ambiguous C-sharp major, a consequence of the revelation of salvation
afforded to Symeon in the person of Christ Himself, a salvation that will be
achieved only by means of the Passion and Resurrection. This moment is, we

may say, a musical symbol. It is something that represents something else, by means of an apophatic understanding of the text. This is very well described by Massimo Cacciari in his short book *Tre icone*, in which he first discusses St. Andrey Rublev's icon of the Trinity and the symbol of the chalice found between the persons of the Trinity represented therein:

> What, then, is it necessary to "cancel" to arrive at such a perfect pres-
> ence of the symbol? The event — the moving into the dimension of the
> *hic et nunc*, the fact as though seized, therefore, during its becoming,
> as momentum. Yes, in that chalice, which is the center of the divine
> Circle, is found the body of the Crucified, but it cannot be to you His
> nature at that moment of the last hour, of the great cry. He is to you
> the form of the Son, "toward" God and God *en arché* [in the begin-
> ning], the end in the Beginning, incarnated-and-resurrected, cruci-
> fied, conqueror of the last enemy. I repeat, a miraculous presence of
> the form. But the presence of the event is "cancelled." Plotinus had
> already said it: the artist who wishes to make the Invisible visible must
> "cancel everything," or, rather: purify his vision in the fire of Sophia
> from all contingency, from all passion.[10]

Purification from passion ensures the understanding of the whole, of the entirety of the history of salvation, rather than the physical reality of the event
that is, as Cacciari said, "cancelled." This is the reason for the use of symbolism: Only thus can we escape from a merely emotional response to what might
seem a human tragedy. This is why, in his *Passio*, Pärt's treatment of the narrative of the Passion of Christ has seemed to many to be cold, or mechanical,
whereas in fact he has sought, precisely as Cacciari suggests, to enter deeply
into the momentum of the conquest of the last enemy, telling the story with
what Wilfrid Mellers described as a "severity of abnegation." This absence is
the only way to ensure presence — in the case of Rublev, of depicting the chalice symbolically formed by the conjunction of the three Persons of the Holy
Trinity. Mellers says of the *Passio*:

> The word is thus rational thought mirroring the ultimate reality of the
> universe which, in being spoken and heard, preserves the distance be-
> tween man and God. In being uttered, the Word becomes a life-giving
> force for men. This amounts to a definition of how art, at least great
> art — and especially music which is the most apparently abstract of the
> arts — functions. The end of Pärt's St John Passion is a revelation of
> this in a very special sense. Only the severity of its abnegation makes it
> possible that, coming to it, we may "have life" thus abundantly.[11]

One might, then, describe the final bars of *Passio* as an aural analogue of the sudden visual discovery of the chalice in Rublev's icon. Because music can only function in time, the listener is obliged to wait for the appearance of the symbol, even though it is also true to say that only by seeing apophatically, so to speak, do we become aware of the symbol that is placed in negative before our very eyes, the chalice right at the center of an icon that might otherwise be construed only as three angels sitting round a table, enjoying the hospitality of Abraham. In *Passio*, Pärt begins with an announcement (the exordium) of the events that will unfold in musical (not to say liturgical) time: "Passio Domini nostri Jesu Christi secundum Joannem," an announcement that takes a mere seven bars and functions as an "up beat" to the unstoppable narrative of the Passion. As Paul Hillier has noted, the resolution of the harmonically ambiguous final chord of this sequence (in which the alto holds a B against what would otherwise be a second inversion of an A minor chord) is the beginning of the Passion narrative itself, in musical terms the A with which the Evangelist initiates the telling of it.[12] The conclusio, lasting for eleven bars, inverts the process, the E pedal of the exordium being resolved, through the descending motion of a scale, onto D, A minor transforming itself into a luminous D major (Figure 3). The whole work is like a pendulum very gradually swinging from one side to another. The end of its trajectory brings us illumination in the simultaneous acknowledgment of the human suffering of Christ ("Qui passus es pro nobis") and the recognition that it is only that suffering that brings redemption ("Miserere nobis"). In other words, it is only when the pendulum reaches the furthest point of its journey that we suddenly become, like the viewer of Rublev's icon, aware of the Chalice.

Such a symbolic approach is also evident in Pärt's *Stabat Mater*, which presents the listener with an aural image of the Mother of God at the foot of the Cross. The sorrow of the Crucifixion, the lamentation of the Theotokos who sees her Son and her God crucified, inevitably contains the joyful essence of the Resurrection, but without the experience of the former we cannot detect or experience the latter. Thus it is that in this work we hear the three solo voices in unison only once, almost at the end of the work — "Quando corpus morietur, fac, ut animae donetur paradisi Gloria." (Figure 4). The only thing necessary thereafter is assent, given by the "Amen" of the three voices in turn. It is through hearing the story of the Passion that we are enabled to understand the Resurrection, and by seeing the Holy Trinity in symbolic form seated round that table that we see the chalice of the Last Supper and realize the connection with the spilled, redeeming blood of Christ. Everything, as Cacciari said, is "cancelled," the artist's purified vision enabling the spectator to see.

Figure 3. *Passio*, "Conclusio." Arvo Pärt "Passio | Passio Domini nostri Jesu Christi secundum Joannem|für Soli, gemischten Chor, Instrumentalquartett und Orgel" © Copyright 1982, 1985 by Universal Edition A.G., Wien/UE17568.

Et resurrexit

Cacciari's "cancelling" and the subsequent enabling of this purified vision means that we see the same things but with different eyes. The change thus effected, anticipated in Pärt's understanding of "verbum" as a turning point, enables an ineffable joy inspired by the Resurrection, without which all would, of course, from the Christian perspective, be in vain. One could plausibly argue, nevertheless, that the composer's vision of the Resurrection is also in a sense apophatic, and the continuous, "dry" text of a work such as *Summa*, a setting of the Creed in Latin, would lend weight to such an argument, as would the setting of the same text in the *Berliner Messe*. Similarly, Wilfrid

Figure 4. *Stabat Mater*, mm. 414–422 Arvo Pärt Pärt "Stabat Mater für Sopran, Countertenor (Alt), Tenor, Violine, Viola und Violoncello" © Copyright 1985 by Universal Edition A.G., Wien/ UE33953.

UE 33953

Mellers described the end of *Passio*, the brief plea "Qui passus es pro nobis, miserere nobis. Amen," not without justice, as "grand, yet pain-ridden,"[13] and it is always without any abnegation of humanity, of the pain that is part of the human condition, that Pärt sees the victory of the Resurrection: One senses always the composer's own description of his tintinnabuli technique as a constant dialogue between the self and forgiveness,[14] so that no vision of Paradise is facilitated at the expense of the remembrance of incarnate, sinful humanity. And, as we have seen, the tearing of the fabric of silence that enables that reentrance of Paradise into human life is effected with extreme gentleness.

Cacciari's "cancelling of everything," a cancelling so paradoxically audible in Pärt, is also clearly foreseen in St. Gregory of Nazianzus. Concerning the Incarnation, he wrote,

> The absence of limit is contemplated in two ways, with regard to the beginning and to the end, for that which is above both and is not contained between them is without limit. When the mind gazes steadfastly into the depth above, not having a place to stand and relying on the representations it has of God, from this perspective it names as "without beginning" that which is without limit and without outlet. Yet when it gazes at what is below and what is subsequent, it names it "immortal" and "indestructible"; and when it views the whole together, "eternal." For eternity is neither time nor some part of time, nor is it measurable, but what is time for us, measured by the movement of the sun, is for everlasting being eternity, since it is coextensive with these beings, as if it were a kind of movement and interval of time.[15]

"What is time for us" is indeed the only means by which we can contemplate the "absence of limit" in this context. The "cancelling of everything" means that we can, paradoxically, gaze into what St. Gregory calls the "depth above," its perspectives numberless but nevertheless channeled through our human perception of the relation between eternity and the mortal measurement of time. Pärt has spoken of this "cancelling" in terms of the objective quality of his text setting, for such a cancelling contains within it the seed of a new flowering. Leopold Brauneiss notes that "the concept of objectivity does point towards an attitude whereby the *ego* is allowed to retreat, not in order to destroy it but to open it to outside influences — in this case to that of language."[16] The "dryness," or "coldness," of the setting of the Latin text in *Summa* and the *Berliner Messe* is therefore an abnegation intended, by an exploration of the text, in a specific language, to go far beyond simple word painting or, indeed, any kind of overtly emotional reaction to it, to envisage the "depth above."

We return to the image of the *Platytera*, the containing of infinity within a space defined by the human condition. In the case of these specific works by Pärt, the reflection and definition of the human condition is symbolized by language, the Latin providing the composer with a ground on which to construct his sonic cathedral, in that, to recall the words of St. Gregory of Nazianzus, "time for us, measured by the movement of the sun, is for everlasting being eternity, since it is coextensive with these beings, as if it were a kind of movement and interval of time"—the Incarnation and its consequences are necessarily measured, envisioned, and sounded by means of our own human limits.

As I have endeavored to demonstrate, Pärt's music is deeply and instinctively engaged with a very long tradition of Christian theological symbolism. The relationship between silence and sound, in particular, manifests an understanding of the creative act as parallel with and representative of the Incarnation and, consequently, the earthly life of Christ. What is more, though Pärt has never been afraid to avail himself of musical techniques of considerable complexity, there is nothing drily intellectual about his approach; it comes from a deep apprehension of the significance of the Incarnation in the daily life of mankind, within which is included the composition—and, vitally, the sounding—of music.

Notes

1. Arvo Pärt, in Kristina Kõrver et al., eds., *In Principio: The Word in Arvo Pärt's Music* (Tallinn: Arvo Pärt Centre 2014), 5.

2. Arvo Pärt, liner notes to *Tabula rasa*, ECM New Series 1275, 1984.

3. See Peter C. Bouteneff, *Arvo Pärt: Out of Silence* (Yonkers, NY: SVS, 2015).

4. Maximos the Confessor, *Selected Writings*, trans. George C. Berthold (New York: Paulist Press, 1985), 120.

5. Romano Guardini, *The Lord* (Washington, DC: Gateway, 1996), 15.

6. Andrew Clements, "Pärt: In Principio; La Sindone; Cecilia Virgine Romana; etc, Estonian Philharmonia Chamber Choir & National SO/Tallinn CO/Kaljuste," *Guardian*, Friday, April 17, 2009, https://www.theguardian.com/music/2009/apr/17/part-in-principio-review.

7. Dale Nelson, "The Bright Sadness of Arvo Pärt," http://www.touchstonemag.com/archives/article.php?id=15-02-042-b.

8. John Behr, *The Mystery of Christ: Life in Death* (Crestwood, NY: SVS, 2006), 16.

9. St. Ephrem of Syria, *Hymns on the Nativity*, trans. K. E. McVey (Mahwah, NJ: Paulist, 1989), 10.6–8.

10. "Ma che cosa è necessario 'cancellare' per pervenire a una tale perfetta presenza del simbolo? L'evento — il volto nella dimensione dell'hic et nunc, il fatto così come còlto nel suo divenire, come momentum. Sì, in quel calice, che è il centro del Cerchio divino, si trova il corpo del Crocefisso, ma non può esservi il suo volto nel momento dell'ultima ora, del grande grido. Vi è la forma del Figlio, 'verso' Dio e Dio 'en arché', fin nell'Inizio, incarnato-e-risorto, crocefisso, vincitore dell'ultimo nemico. Miracolosa presenza della forma, ripetiamo. Ma la presenza dell'evento è 'cancellata'. Plotino lo aveva detto: l'artista che voglia rendere visibile l'Invisibile dovrà 'cancellare tutto', ovvero: purificare al fuoco della Sophia la sua visione da ogni contingenza, da ogni passione." Massimo Cacciari, *Tre icone* (Milano: Adelphi, 2007), 23–24.

11. Wilfrid Mellers, "Arvo Pärt, God, and Gospel: Passio Domini Nostri Iesu Christi Secundum Iohannem" [1982], *Contemporary Music Review* 12, no. 2 (1995): 48.

12. Paul Hillier, *Arvo Pärt* (Oxford: Oxford University Press, 1997), 129.

13. Mellers, "Arvo Pärt, God, and Gospel," 40.

14. Hillier, *Arvo Pärt*, 96.

15. St. Gregory of Nazianzus, "Oration 38: On the Nativity of Christ," in *Festal Orations*, trans. Nonna Verna Harrison (Crestwood, NY: SVS, 2008), 66.

16. Leopold Brauneiss, "Tintinnabuli: An Introduction," in *Arvo Pärt in Conversation*, ed. Enzo Restagno et al. (Champaign, IL: Dalkey Archive, 2012), 125.

13

Christian Liturgical Chant and the Musical Reorientation of Arvo Pärt

Alexander Lingas

Arvo Pärt has often said that Gregorian chant played a seminal role in the process of artistic and spiritual reorientation that followed *Credo* (1968), his last major essay in musical collage, and his consolidation of a new compositional approach in a series of works presented under the rubric of "tintinnabuli" in a concert held in Tallinn on October 27, 1976.[1] The composer and his wife have also described these eight years as a period of near-total compositional "silence," during which Arvo sought to overcome a creative impasse through immersing himself in sacred monophony and polyphony from the premodern West. Out of this emerged the revelation of "tintinnabuli," a compositional technique founded on a two-part texture in which a melodic "M-voice" is accompanied by a "T-voice" arpeggiating a triad built on the home note of the governing tonality.[2] Pärt has strengthened the narrative of a mainly silent period of study by retaining in his catalogue of authorized music only a single large-scale work from this time, namely, Symphony No. 3 (1971), the polyphonic writing and archaic cadences of which bear witness to his interest in medieval and Renaissance music.

In his pioneering 1997 monograph, Paul Hillier accepted 1968–1976 as a period of compositional silence, even as he acknowledged that "Pärt wrote reams of stylistic and technical studies resulting from his exploration of early music; he also continued to support himself and his family by writing film music, though only until 1974."[3] Claiming "that in the eyes (or ears) of most people, it is the influence of early music that has had the most readily discernible impact on his style," Hillier then attempted in some detail to demonstrate how Pärt's tintinnabuli music was indeed "rooted in plainsong" by indicating elements it held in common with medieval Latin plainchant and organum, including procedures for its generation from sacred text, evocations of responsorial chanting, and use of drones.[4] Peter Phillips questioned the strength of these associations in a sarcastically titled review ("Holy Minimalism!"), remarking,

> Apart from the use of a drone, and the coincident employment of the same notation for the two, there is little apparent connection between the quoted examples of Pärt's music and the early music on which it is supposed to be based. This is just one example among many of forcing too much on Pärt's writing. The more obvious truth, for better or worse, is that Pärt's style is a style fashioned for modern use.[5]

One may agree with Phillips that tintinnabuli is inescapably a product of our own age without, however, severing its links to plainsong. Although Pärt has rarely quoted historic chant melodies in his tintinnabuli works,[6] many of their professional and lay listeners have perceived them as somehow extending pre-existing traditions of monophonic sacred song, especially those of Gregorian chant. Musicologists, for example, have often noted stylistic parallels between the M-voices of tintinnabuli works and plainsong melodies,[7] and Paul Hillier has counseled performers that experience with plainchant is invaluable for rendering Pärt's vocal music.[8] Meanwhile in popular discourse the sound and spiritual ethos of his music are regularly conflated with those of Gregorian chant as now commonly understood, an identification that has contributed to the highly problematic placement of Pärt's tintinnabuli works within a category of "holy minimalism."[9] Finally, there is the sheer consistency with which the composer has insisted that Gregorian chant played a significant role in starting and then guiding his journey toward tintinnabuli.

Working in the spirit of recent scholarship that has enriched our understanding of change and continuity leading up to, across, and beyond Pärt's nominally "silent" period of 1968–1976,[10] I will devote the main body of this chapter to reconsidering his relationship to Christian plainchant. Recollections by Arvo and Nora Pärt will provide context for a preliminary report on a large body of monophony that the composer wrote as musical and devotional exercises in eleven notebooks dated from February to November 1976, a period during which the first works in the tintinnabuli style were completed and publicly performed.[11] Some brief observations of the ways these melodies are, despite Pärt's intense study of Gregorian chant, *unlike* those found in historic repertories of Christian liturgical song will lead me to identify some ways that the tintinnabuli works that follow them chronologically manifest congruencies with plainchant. My aim in so doing is not to compile a list of musical features borrowed consciously or unconsciously from the traditions of Latin, Byzantine, or Slavonic liturgical singing (many of which have been noted previously by other scholars) but to suggest that in tintinnabuli emerged a musical system that in some ways recreates by different means some of their organizational procedures.

Pärt as a Student and Composer of Chant

Pärt is unsure of exactly when and where it was that he heard the recording of Gregorian chant that suggested that a route out of his creative impasse might be sought in the music of the distant past,[12] but in interviews since the late 1980s he has not wavered in his conviction that the encounter was for him decisive. Speaking in 2000 to Jordi Savall, he recalled:

> All I know is that when I heard Gregorian chant for the first time, I must have been mature enough, in one way or another, to be able to appreciate such musical richness. At that moment I felt at once utterly deprived and rich. Utterly naked, too. I felt like the prodigal son returning to his father's home. I had nothing, I had accomplished nothing. The methods I had used before had not allowed me to say what I wanted to say with music, yet I did not know any others. At that moment, my previous work seemed like an attempt to carry water in a sieve. I was absolutely certain: everything I had done until then I would never do again. For several years I had made various attempts to compose using collage techniques, mainly with the music of Bach. But all of that was more a sort of compromise than something I carried in my flesh. Then this encounter with Gregorian music . . . I had to start again from scratch. It took me seven, eight years before I felt the least bit of confidence — a period during which I listened to and studied a lot of early music, of course.[13]

In likening himself the Prodigal Son of the biblical parable, Pärt was referring obliquely to the inseparability of his musical and spiritual motivations for immersing himself in music of the premodern West.[14] Arvo and Nora Pärt recall that he studied not only Gregorian chant, for which a copy of the *Liber usualis* was his main resource,[15] but also polyphonic music, including that of "the Notre Dame school, then Machaut, Franco-Flemish music, Obrecht, Ockeghem, and Josquin," eventually supplemented by later Renaissance masters including Palestrina and, especially, Victoria.[16] Influence of the old polyphonic masters soon surfaced in Symphony No. 3, a work that left Pärt "dissatisfied" with being so obviously indebted to musical history but "encouraged [him] to go on exploring."[17] Central to this exploration were efforts to divine the "cosmic secret . . . hidden in the art of combining two, three notes" that gave early music, especially Gregorian chant, the "life-giving power" that he instinctively felt it possessed.[18] To accomplish this, he complemented his study of historic chants with the composition of a substantial body of monophony that he recorded in the notebooks that serve him as a chronicle of his musical and nonmusical thoughts.[19]

The eleven notebooks dating from February to November 1976 that I examined in January 2018 show that Pärt was still intensely engaged in the writing of monophonic exercises even as he was forging the techniques of tintinnabuli. Although time did not permit a full inventory, my impression was that the number of folios Pärt devoted to monophony in them exceeds that allotted to sketches composed in multiple parts or presented in graphic form.[20] The indivisibility of music and religion for Pärt is evident from the fact that devotional texts account for approximately one-fourth of the material recorded in these notebooks. Alongside texts in Estonian and a few in Latin one finds a significant body of writing in Russian and Church Slavonic, including prayers, biblical citations, and passages from spiritual authorities ranging chronologically from Athanasius (fourth century) to Seraphim of Sarov (1754–1833).[21]

Pärt's melodic experiments are for the most part wordless, with the exceptions including a kyrie from the Mass ordinary of the Roman rite[22] and a set of sketches setting the Jesus Prayer in Slavonic. The latter begin with recitation on a single pitch (E) followed by more tuneful renderings, some of which feature melodic extensions of certain syllables (melismas), with melodic finals on G, F-sharp, and E.[23] Although many of Pärt's exercises employ key signatures with one or two sharps, his melodies are mainly diatonic and more tonal than modal, with a preference for minor keys inflected with occasional leading tones. Their rhythmic profiles show considerable variety yet are essentially of two main types, either Solesmes-style "free rhythm" or ternary patterns recalling the rhythmically measured (mensural) chant and early polyphony of the medieval Latin West. There is a similar variety of melodic contours and ranges, with some tunes moving predominantly in stepwise motion over a narrow range and others featuring many leaps roving over a wide ambitus.

Notebooks 8, 9, and 10 contain Pärt's most ambitious exercise in monophony: an effort accomplished between April and June 1976 to compose in sequence a melody corresponding to each of the 150 psalms in the Bible. Nora Pärt has identified this experiment as one of a number of attempts by her husband at that time to "develop his spontaneity," attempts that also included writing melodies prompted by sketches he made upon observing "flocks of birds."[24] The composer has described his working method and its results as follows:

I read a psalm and filled a whole page, without thinking much about it, in the hope that there might be some sort of connection between what I wrote. Then I got on with reading the next psalm. I'm sure that between the psalm and my melody there was no connection at all. At least I didn't notice any. However, I still hoped for some sort of osmosis, and so I went on and repeated the same process with one hundred and fifty psalms.[25]

Unlike Pärt's later tintinnabuli settings of psalms, there is indeed no evident correlation in length between these exercises in monody and the texts of their corresponding psalms, which are indicated in the notebooks by superscriptions. These superscriptions move in sequence through the Psalter and are normally followed by a single chant, but in some cases multiple chants or other kinds of sketches are recorded before the next psalm heading. The melodies that arose from this extended experiment are predominantly notated in E minor and in free rhythm, although some chants feature short or extended mensural passages.[26] They exhibit the full range of conjunct and disjunct melodic contours evident in his other essays in monody from 1976, yet they tend toward extreme ranges and frequent juxtapositions of stepwise and angular writing.

Pärt has voiced deep dissatisfaction with the body of monody that he composed before consolidating his tintinnabuli style, going so far as to describe it in interviews as "mad things," "dead exercises," and "absolutely meaningless things."[27] Looking upon these melodies from the perspective of a scholar and performer of historical and living traditions of Christian plainchant, I was forced to conclude that he was for the most part right about their musical value. Other than a few items that stylistically resemble traditional chants,[28] Pärt's melodies do not demonstrate a musical logic comparable to that evident to listeners and analysts in the central repertories of notated Byzantine, Latin, and Slavonic liturgical chant. His chants, especially the psalm exercises, also fare badly when judged against the products of spontaneous composition within living Arab, Persian, and Indian traditions of monophonic music. Lacking the musical grammar provided by the modal systems of these traditions, Pärt wrote melodies that ventured capriciously into extreme registers without adequate preparation or development. Not unexpectedly for a composer reared in tonal harmony, in which "chords become decisive co-determinants" for melody,[29] he evidently did not know how logically to develop and sustain musical interest in a single line in the manner of either melismatic Christian chant or an instrumental *taksim*. Bearing this in mind, one may understand better why Nora Pärt reports that the composition of these exercises was marked by suffering, for her husband seems to have been unable to master through the medium of monophony the "cosmic secret" he had perceived in Gregorian chant "hidden in the art of combining two, three notes."[30]

Concluding Reflections on Tintinnabuli and Christian Plainchant

Having reached an impasse in his explorations of monophony, Pärt "discovered and developed" in tintinnabuli, as Brauneiss has observed, a compositional means "to recreate the effect which Gregorian [chant] is capable of producing on the listener of today."[31] Based in part on a quarter-century of experience with performers and audiences as musical director of the vocal ensemble Cappella Romana, however, I would broaden this perception of commonality to extend beyond Gregorian chant to embrace other ancient forms of Latin liturgical song, Byzantine and Slavonic chant, and some repertories of premodern polyphony. Significantly, for some of us cognizance of these affinities seems to have been independent of the profound sonic differences between renderings of Pärt's tintinnabuli vocal works by voices trained in the English or Scandinavian choral traditions (a stylistic norm promulgated by their "authorized" recordings on the ECM label) and microtonally inflected renderings of chant rooted in received traditions of Byzantine performance practice. By way of a conclusion, I would like to indicate briefly how Pärt, informed by his study of the monophony and early polyphony of the premodern West yet unwilling simply to recapitulate what his wife called "cold, dead rules from years gone by,"[32] ended up creating in tintinnabuli a musical system that grounded his melodic imagination in procedures that were in some respects congruent with those governing historic traditions of Christian plainchant.[33]

Nora Pärt has observed that the fundamental two-part texture of tintinnabulation provided her husband with a new way to regulate musical tension, something that we have noted proved difficult for him when writing bare monody. Analysts have subsequently developed her insight, demonstrating how Pärt alternates tension and release through such techniques as the variation of intervallic density, evident in changes to the spacing of chords and the proportion of dissonant intervals.[34] Yet it is worth recalling the Pärts' insistence that in tintinnabuli the melodic M-voice and accompanying arpeggiation of the tonic in the T-voice should also be considered as grounded in a single reality, so that 1+1=1.[35] John Roeder confirmed the underlying unity of tintinnabulating voices when he showed how "Pärt's strict system makes harmony into an aspect of melody."[36] Roeder's conclusion in turn recalls the Aristoxenian conception of harmony "as a formative participant of monophonic music,"[37] a theoretical notion inherited by the monophonic traditions of the late antique and medieval Mediterranean, including Christian plainchant. William Thomson has provided a way of understanding the acoustical and cognitive reasons for such functional equivalence in a cross-cultural study of how melody is governed by "tonality frames," which he defines as "a conception of organization that ties the spectral content of the single tone to the chord and to melody, both bearing in common the properties of sonance and root."[38]

Viewed from this perspective, tintinnabuli provided Pärt with a way to discipline his melodic imagination that, while retaining some elements of Western traditions of tonal polyphony and harmony,[39] in other respects resembles plainchant and certain other ancient traditions of monophony. These resemblances include:

1. The setting of melodies within areas of tonal stability confirmed by the melodic gravity exerted by modal finals that, in some traditions such as Byzantine chant, are made audible as drones. In Pärt's tintinnabuli music the work of sounding the prevailing tonality is usually accomplished by one or more T-voices that in some cases are supplemented or supplanted by an actual drone.[40]

2. The use of logogenic procedures for the generation of melody, the strict application of which in such early tintinnabuli works as *Missa syllabica* and *Passio* served to create austere stepwise vocal lines that constantly returned to their tonal homes. Encapsulated by the composer in the phrase "the words write my music," these techniques serve as comparatively simpler substitutes for the mechanisms governing successions of musical formulas and their application to text evident in musically developed genres of Christian plainchant.[41] Additional technical parallels between Pärt's melodic writing and liturgical chant have been noted in his creation of melismatic chant in the *Te Deum* by interpolating T-voice formulas within stepwise logogenic lines and a fondness for rhetorical gestures that becomes more pronounced as he develops the expressive potential of tintinnabuli.[42]

3. The creation of large-scale structures from combinations of techniques held in common with chant, especially that belonging to highly melismatic genres in which musical considerations eclipse text in the structuring of form, with others particular to tintinnabuli, notable among which is the regulation of levels of dissonance between voices. In contrast to the goal-directed tonal structures that Karol Berger has identified as characteristic of Western art music from the later eighteenth century onward,[43] one finds both in works by Pärt and, for example, repertories of melodically florid Byzantine chant the operation of what Cizmic has described as "process-driven stasis."[44] In such music, moments of tension and release are

certainly perceived but serve ultimately to emphasize the stability or even identity of the points of musical arrival and departure in ways that highlight contrasts between human time and the eschatological experience of the perpetual heavenly liturgy.[45]

Having reached the end of this chapter, there is neither space nor need, thanks to the work of other scholars, to append here a catalogue of minor congruencies between Pärt's published music and repertories of Christian liturgical chant, especially Eastern Orthodox ones, that have proliferated since the 1990s.[46] Instead I shall simply convey some additional information regarding Pärt as a composer of chant that he personally brought to my attention during my January 2018 visit to the Arvo Pärt Centre. As I was examining his notebooks from 1976, he suggested that I also look at four notebooks from the summer of 1991 containing entries written in multiple locations — Berlin, Ghent, Rome, Stockholm, Tolleshunt Knights (UK), and Zürich — that reflected his ascent to international prominence.[47] Their contents showed that he continued to use notebooks as a means to chart his musical and devotional life with quotations from spiritual authorities, the recording of personal observations, and the composition of musical sketches that include numerous instances of monophony displaying the familiar mix of freely and mensurally notated rhythms. Textless melodies abound, but so do settings of the Jesus Prayer and the exclamation "Lord, have mercy" in Slavonic, Estonian, and even Greek. Perhaps the most significant difference between the monophony he composed in 1976 and that written in 1991 is that the newer melodies more closely resemble traditional Christian chant, a style that, as we have seen, he came to approximate in his public works thanks to the strictures of tintinnabulation.

Notes

1. Notably Martin Elste, "An Interview with Arvo Pärt," *Fanfare* 11 (March/April 1988), https://www.arvopart.ee/en/arvo-part/article/an-interview-with-arvo-part/; Roman Brotbeck and Roland Wächter, "Lernen, die Stille zu hören: Ein Gespräch mit dem estnischen Komponisten Arvo Pärt," *Neue Zeitschrift für Musik* 151, no. 3 (1990): 14–15; Enzo Restagno et al., *Arvo Pärt in Conversation*, trans. Robert J. Crow (Champaign, IL: Dalkey Archive, 2012), 18, 28; Arthur Lublow, "The Sound of Spirit," *New York Times*, October 15, 2010; Jordi Savall, "A Conversation with Arvo Pärt. Translated from the French by Taylor Davis-Van Atta and Katherine Linton," *Music and Literature* 1 (2012): 7. Pärt also describes his formative encounter with Gregorian chant in Dorian Supin's film *24 Preludes for a Fugue*, a scene discussed in Laura Dolp, "Arvo Pärt in the Marketplace," in *The Cambridge Companion to Arvo Pärt*, ed. Andrew Shenton (Cambridge: Cambridge University Press, 2012), 183;

Immo Mihkelson, "A Narrow Path to the Truth: Arvo Pärt and the 1960s and 1970s in Soviet Estonia," in *The Cambridge Companion to Arvo Pärt*, ed. Andrew Shenton (Cambridge: Cambridge University Press, 2012), 27.

2. This nomenclature, now standard in the field, is from Paul Hillier, *Arvo Pärt* (Oxford: Oxford University Press, 1997), 92–93.

3. Hillier, *Arvo Pärt*, 66.

4. Hillier, *Arvo Pärt*, 77–85.

5. Peter Phillips, "Holy Minimalism!" *New Republic*, December 1, 1997, 52.

6. Exceptions are the introit *Statuit ei Dominus*, which frames tintinnabuli music with quotations of the Gregorian version, and *Da pacem*, which features the melody of the Gregorian antiphon as a cantus firmus in the alto voice. In addition to the Pärt's discussion of the former in Restagno et al., *Pärt in Conversation*, 28, see Andrew Shenton, *Arvo Pärt's Resonant Texts: Choral and Organ Music, 1956–2015* (Cambridge: Cambridge University Press, 2018), 150–51, 236–38.

7. For example, Leopold Brauneiss, "Tintinnabuli: An Introduction," in *Arvo Pärt in Conversation*, ed. Enzo Restagno et al. (Champaign, IL: Dalkey Archive, 2012), 124; Constantin Gröhn, *Dieter Schnebel und Arvo Pärt: Komponisten als "Theologen"* (Münster: Lit, 2006), 159; Oliver Kautny, *Arvo Pärt Zwischen Ost und West: Rezeptionsgeschichte* (Stuttgart: Metzler, 2002), 93, 214, 23, 46–49; Olga Valerievna Osetskaia, "Священное Слово В Музыке А. Пярта," Kandidat [PhD] diss., Нижегородская Государственная Консерватория им. М.И. Глинки, 2002, 100–2; Svetlana Savenko, "Musica Sacra of Arvo Pärt," in *"Ex Oriente . . .": Ten Composers from the Former USSR: Viktor Suslin, Dmitry Smirnov, Arvo Pärt, Yury Kasparov, Galina Ustvolskaya, Nikolai Sidelnikov, Elena Firsova, Vladimir Martynov, Andrei Eshpai, Boris Chaikovsky*, ed. Valeria Tsenova (Berlin: Ernst Kuhn, 2002), 165; Shenton, *Pärt's Resonant Texts*, 118, 36–99.

8. Paul Hillier, "Some Observations on the Performance of Arvo Pärt's Choral Music," in *The Ashgate Research Companion to Minimalist and Postminimalist Music*, ed. Keith Potter and Kyle Gann (Farnham: Ashgate, 2013), 390.

9. Kautny, *Pärt Zwischen Ost und West*, 161–62, 246–62.

10. Peter C. Bouteneff, for example, has shed light on the theological and devotional sources that inspired the composer during these eight years (and beyond) in *Arvo Pärt: Out of Silence* (Yonkers, NY: SVS, 2015). Other scholars have revealed aspects of continuity across Pärt's career, even to the point of reevaluating the extent to which 1968 to 1976 was actually a period of compositional silence. Particularly notable in this regard are two doctoral theses: Mark Eric John Vuorinen, "Arvo Pärt's Serial and Tintinnabuli Works: A Continuum of Process," DMA, University of Toronto (Canada), 2014; Christopher Jonathan May, "System, Gesture, Rhetoric: Contexts for Rethinking Tinitnnabuli in the Music of Arvo Pärt, 1960–1990," D.Phil. thesis, University of Oxford, 2016.

11. Varying in size from eighteen to eighty-five folios, these are notebooks No. 5 (4–29 February 1976 [the year was changed on the cover to 1975, but Toomas Schvak has confirmed that 1976 is correct]), No. 6 (1–31 March 1976), No. 7 (1–15 April

1976), No. 8 (16 April–20 May 1976), No. 9 (20–25 May 1976), No. 10 (26 May–5 June 1976), No. 11 (6–12 June 1976), No. 13 (1–16 July 1976), No. 14 (17–31 July 1976), and No. 16 (3 September–27 November 1976). Copies were graciously made available to me for study at the Arvo Pärt Centre in Laulasmaa, Estonia, on January 15–16, 2018. My thanks to Toomas Schvak for kindly fielding queries from afar and, especially, to Kristina Kõrver for her extraordinary efforts to ensure that my short visit was as enjoyable and productive as possible. I am also grateful to Arvo Pärt, for suggesting that I look at the 1991 notebooks, and to Nora Pärt, for a wide-ranging conversation conducted, with the assistance of Ms. Kõrver, in multiple languages.

12. Agreeing with one interviewer that the encounter happened around 1970 (Savall, "A Conversation," 8), Pärt has described it as occurring in a record store (Brotbeck and Wächter, "Lernen," 18; Restagno et al., *Pärt in Conversation*, 18) or bookstore (Supin, *24 Preludes*), although it is possible that he was listening to a radio broadcast (Lubow, "The Sound of Spirit").

13. Savall, "A Conversation," 7–8.

14. Savall, "A Conversation," 8. See also Brotbeck and Wächter, "Lernen," 15; Lublow, "The Sound of Spirit"; and Jamie McCarthy, "An Interview with Arvo Pärt," *Contemporary Music Review* 12, no. 2 (1995): 59.

15. Brotbeck and Wächter, "Lernen," 15; Restagno et al., *Pärt in Conversation*, 28.

16. McCarthy, "An Interview," 59; Restagno et al., *Pärt in Conversation*, 33–35.

17. Restagno et al., *Pärt in Conversation*, 29.

18. Elste, "An Interview with Arvo Pärt"; Lublow, "The Sound of Spirit."

19. Restagno et al., *Pärt in Conversation*, 29. It is important not to conflate Pärt's study of Gregorian chant with his compositional exercises in monophony and speak of him as "copying out" psalms from the *Liber usualis* (Vuorinen, "Arvo Pärt's Serial and Tintinnabuli," 60) or "fill[ing] his notebooks with ancient melodies" (Lublow, "The Sound of Spirit").

20. In some cases (for example, folios 23r–30v of Notebook 5) the dividing line between monophonic exercises to which additional voices have intermittently been added and sketches conceived at the outset as being in multiple parts is not always clear.

21. For example, respectively, Notebook 8, f. 2v and 42r.

22. Notebook 6, f. 22r–22v, including some passages with the addition of a second voice.

23. Notebook 10, f. 29r–44v. Additional musical settings of the Jesus Prayer occur in Notebooks 13, 14, and 16.

24. Restagno et al., *Pärt in Conversation*, 29. See also Savall, "A Conversation," 13–14.

25. Restagno et al., *Pärt in Conversation*, 29.

26. For example, the melodies corresponding to Psalms 49 and 50 both feature brief mensural passages (Notebook 9, f. 49v–50r).

27. Savall, "A Conversation," 13–14. The composer described this music to me in similar terms during my January 2018 visit to Laulasmaa. Elsewhere he had also been

unhappy, albeit primarily for timbral reasons, when members of Andres Mustonen's early music ensemble Hortus Musicus played some of these "melodies on the violin or crumhorn, particularly with the latter instrument." Restagno et al., *Pärt in Conversation*, 32–33.

28. For example, a melody in F major dated 29 February 1976 (Notebook 5, f. 38r) in the style of a Gregorian antiphon.

29. William Thomson, "From Sounds to Music: The Contextualizations of Pitch," *Music Perception: An Interdisciplinary Journal* 21, no. 3 (2004): 449.

30. Restagno et al., *Pärt in Conversation*, 30; Elste, "An Interview with Arvo Pärt."

31. Brauneiss, "Tintinnabuli: An Introduction," 117.

32. Restagno et al., *Pärt in Conversation*, 29.

33. Cf. Sean McClowry's observation that Pärt "was interested in returning back to the origins of classical music to reinvent new alternatives to the conventions that were set in motion at this important crossroads in music history." Sean McClowry, "The Song of the Convert: Religious Conversion and Its Impact on the Music of Franz Liszt, Arvo Pärt, and John Coltrane," PhD diss., Princeton University, 2012, 43–44; and Shenton, *Pärt's Resonant Texts*, 34.

34. For example, Carol Leonore Matthews Whiteman, "*Passio*: The Iconography of Arvo Pärt," PhD diss., City University of New York, 1997, 48–117; and Shenton, *Pärt's Resonant Texts*, 43–46.

35. Restagno et al., *Pärt in Conversation*, 30–31.

36. John Roeder, "Transformational Aspects of Arvo Pärt's Tintinnabuli Music," *Journal of Music Theory* 55, no. 1 (2011).

37. Thomson, "From Sounds to Music," 442.

38. Thomson, "From Sounds to Music," 431.

39. On the retention of common practice harmony in Pärt's music, see Shenton, *Pärt's Resonant Texts*, 37, 42.

40. During our conversation in Laulasmaa, Nora Pärt endorsed the idea that the T-voice of tintinnabulation functions musically in a way comparable to the Byzantine tradition of *isokratema* (drone holding). Other scholars who have followed on from Pärt's labeling of the recorded drones in his *Te Deum* as "ison" to discuss more broadly the question of affinities between Byzantine *isokratema* and tintinnabuli include Osetskaia, "Священное Слово В Музыке А. Пярта," 104; Savenko, "Musica Sacra," 165; Shenton, *Pärt's Resonant Texts*, 86, 136–41.

41. Authors who have remarked on parallels between centonization in liturgical chant and Pärt's logogenic melodic writing are Gröhn, *Dieter Schnebel und Arvo Pärt*, 159; Osetskaia, "Священное Слово В Музыке А. Пярта," 100–1; and Nora Potemkina, "Современное Православное Богослужебное Пение И Духовная Музыка Рубежа Xx–Xxi Веков. Параллели И Пересечения," *Ученые записки РАМ имени Гнесиных* 3, no. 18 (2016). The literature on logogenic composition and centonization in Christian plainchant is vast. Two examples that discuss chant of the Roman rite are William T. Flynn, *Medieval Music as Medieval Exegesis* (Lanham,

MD: Scarecrow, 1999); and Emma Hornby, *Medieval Liturgical Chant and Patristic Exegesis: Words and Music in the Second-Mode Tracts* (Woodbridge: Boydell, 2009).

42. Osetskaia, "Священное Слово В Музыке А. Пярта," 81–88; Shenton, *Pärt's Resonant Texts*, 136–41, 97.

43. Karol Berger, *Bach's Cycle, Mozart's Arrow: An Essay on the Origins of Musical Modernity* (Berkeley: University of California Press, 2007).

44. Maria Cizmic, "Transcending the Icon: Spirituality and Postmodernism in Arvo Pärt's *Tabula Rasa* and *Spiegel im Spiegel*," *Twentieth-Century Music* 5, no. 1 (2008): 75. I discuss examples of such chants from Byzantium in Alexander Lingas, "From Earth to Heaven: The Changing Soundscape of Byzantine Liturgy," in *Experiencing Byzantium: Papers from the 44th Spring Symposium of Byzantine Studies*, ed. Claire Nesbitt and Mark Jackson (Aldershot: Ashgate, 2013).

45. Pärt speaks of eternal nature of angelic music in Brotbeck and Wächter, "Lernen," 16. These ideas are explored further in Kevin C. Karnes, *Arvo Pärt's Tabula Rasa* (New York: Oxford University Press, 2017), 52–55; Gröhn, *Dieter Schnebel und Arvo Pärt*, 130; Patrick Revol, *Conception orientale du temps dans la musique occidentale du vingtième siècle* (Paris: Harmattan, 2007), 182–84; Shenton, *Pärt's Resonant Texts*, 287; and Brunhilde Sonntag, "Das Problem der Zeit in der Tintinnabuli-Musik Arvo Pärts," in *Arvo Pärt — Rezeption und Wirkung seiner Musik*, ed. Oliver Kautny (Osnabrück: Electronic Publishing, 2001), 41–44.

46. See, for example, Oliver Kautny, "«Alle Länder, in Ihrer Verschiedenheit, Sind Eins . . .» : Der Schein des Orientalischen bei Arvo Pärt," *Neue Zeitschrift für Musik (1991–)* 164, no. 2 (2003): 24–27; and Shenton, *Pärt's Resonant Texts*, 151–53, 215, 25.

47. No. 8 (June 1991), No. 9 (July 1991), No. 11 (July–August 1991), and No. 12 (August 1991).

14

In the Beginning There Was Sound

Hearing, Tintinnabuli, and Musical Meaning in Sufism

Sevin Huriye Yaraman

There is a candle in the heart of man, waiting to be kindled.
In separation from the Friend, there is a cut waiting to be stitched.
O, you who are ignorant of endurance and the burning fire of love —
Love comes of its own free will, it can't be learned in any school.
— MEVLÂNÂ MUHAMMED CELÂLEDDIN-I RÛMI

In this chapter I have two distinct but ultimately related objectives. The first is to identify two seemingly oppositional states that substantiate human existence and underlie the theology of Sufism: separation from God in longing and Union with Him in joy. The second is to explore musical meaning within the theology of Sufism. For this second objective I will draw on a variety of textual and musical evidence, seeking to establish the relationship between the state of longing, which marks life in this world, and the power of music as a vehicle for both the remembrance of God and the conveyance of the sorrow of separation from Him.

Both of these areas will be explored with reference to the ethos and underlying sense of Arvo Pärt's "tintinnabuli" concept, notably its way of holding together duality and unity. In his book *Arvo Pärt: Out of Silence*, Peter Bouteneff writes, "The unity of dualities that concerns Pärt's compositional inspiration and processes are essential to the biblical and theological tradition of the Orthodox Church." Therefore, "musical understanding and analysis of Pärt's music," which is said to combine seemingly paradoxical yet unified sensibilities such as sorrow and joy, Bouteneff adds, "will gain a deeper meaning through correlation with theological principles."[1]

With that understanding in mind, this essay will unfold in three sections:

I shall begin by approaching tintinnabuli as an inspiration and a tool to clarify the unity/oneness of the states of binary oppositions in Sufi theology— separation and union. In the second section, I shall turn to the Qur'an and classical Sufi writings in an attempt to locate a theory of longing, the significance of hearing, and musical meaning. Finally, in the third section, to deepen the understanding of the unity of dualities as a fundamental component in Sufism, I shall focus on music and poetry and discuss a Sufi song, "Dolap, niçin inlersin?," through the lens of the expression of longing and the anticipation of joy in reunion.

Tintinnabuli, Duality, and Unity

In the most general terms, tintinnabuli—Pärt's compositional style that reflects the composer's mystical and musical experience with Western sacred chant—is a counterpoint between two distinct musical lines: the melody (M) voice uses notes from the tonic, and the tintinnabuli (T) voice arpeggiates notes from the tonic triad against the M voice. Pärt clarifies the specific symbolic significance of each of the voices, saying that the M-voice signifies the earthly, terrestrial existence, while the T-voice signifies heavenly existence— and calls them "the eternal dualism."[2] At the same time, Pärt explains that "tintinnabuli is the mathematically exact connection from one line to another . . . tintinnabuli is the rule where the melody and the accompaniment [accompanying voice] is one. One and one, it is one—it is not two. This is the secret of this technique."[3] Tintinnabuli's separate yet admittedly unified lines have received several interpretations in an attempt to explicate the dependency of existence and the meaning of human and divine. I distinguish one interpretation from the others for its simplicity and reflection of a fundamental theological principle in Sufism most closely: the multiplicity of one; it is apparently multiple but in the essence is One. This is expressed in the well-known equation devised by the composer's wife Nora Pärt: $1+1=1$. Pärt embraced the equation as an expression of a "heart-rending union," one that the soul yearns to sing endlessly: "one perfect thing."[4] The interaction of two voices in tintinnabuli often creates an audible dissonance, which works seamlessly yet painfully.

Here, in order to make my point, I will suggest a variation on Nora Pärt's equation. This variation offers an analogy to the initial oneness of humanity with God, followed by a painful separation and the existence of multiplicity in the terrestrial life felt and expressed through longing for Him. My variation would be $1=1+1$. That is to say, oneness and union are followed by the separation and multiplicity of 1. However, this is not the end; the painful yet beautiful dissonance ends with a final resolution in life by returning to the

Divine, where sorrowful longing ends and joy begins. Therefore, bringing the final 1 back to revise my proposed equation is now inevitable: The final equation will be 1=1+1=1. Initial oneness and joy = separated and multiplied in this life in sorrow and longing for what is lost = return, oneness. This way of construing essential duality and unity, together with Pärt's understanding of "bright sadness," will now be brought to bear on my reflections on aspects of Sufi theology.

Longing and Yearning in the Qur'an and Sufi Texts in Relation to Musical Meaning

Sufism manifests its theory of longing and music through the idea presented in scripture that the very first interaction between God and every soul was conversation (Qur'an 7:172). God gave all souls the faculties of hearing and speaking and then spoke to them. He began the dialogue by asking the primordial question *Alastu bi-rabbikum*, "Am I not your Lord?" and the souls comprehended and responded: "Indeed we bear witness that You are," attesting to an Eternal Covenant that bound all of humanity to its creator.[5]

The classical theology of Sufism is fundamentally concerned with the state of separation and longing for God; an intense emotion, it is believed, that consumes a person's life. What, then, are the signs and conditions of longing? To answer this question, we will turn to Abu Hamid al-Ghazali, a towering figure of twelfth-century mystical philosophy and theology. In his definitive work *Love, Longing, Intimacy, and Contentment*, al-Ghazali explains three related conditions for longing.[6] It should be noted that al-Ghazali affirms that all our longing reflects the most fundamental one: the yearning for God.

The first necessary condition of longing is love, since only a beloved in "his concealment would become an object of longing. Affirmation of love should suffice for the affirmation of longing."[7] The expression of love and the expression of longing for the beloved are not separate: The more one loves, the more one longs for the beloved.

The second condition is unattainability: The lover must not be able to attain the longed-for beloved. Al-Ghazali clarifies by saying that "a beloved who is present and attained cannot be longed for because longing seeks, and yearns to seek, for something, whereas that which is already found is not sought."[8]

This brings us to al-Ghazali's last condition of longing, which is the most significant for our discussion of musical meaning. The subject of longing has to be at least partially known to the person who yearns, while remaining imperceptible in another aspect, because the utterly imperceptible cannot be a subject of longing. "A person whose beloved is absent," al-Ghazali says, "but in whose heart a knowledge and the remembrance of the loved one remains, longs to complete the image of the loved one. . . . His longing denotes an inner yearning to perfect his imagination."[9] The implication is that longing requires a piece of knowledge that is always remembered, a knowledge that gives the pleasure and the certainty of the beloved's existence, since an entirely unknown object can neither be loved nor longed for.

Islamic theology teaches that the yearning is quenched in the next world by "encountering" God's beauty and witnessing Him with spiritual insight. In other words, the reunion at the end will include the witnessing of God's beauty. But if witnessing is needed to complete the image of God for the seeker, what aspect of God were the souls allowed to acquire as a knowledge in pre-eternity, during the first stage of union, which has been sustained through pain and intense longing? In short, what constitutes that partial, incomplete knowledge that causes human beings to yearn for God during the state of separation on earth?

Islamic theology tells us that it is the voice of God implanted in pre-eternity in each human soul. The sweetness and pleasure of God's voice remains in the soul for all time — His voice, soundless and wordless, by which the creation of the universe receives vibration and becomes sound to be heard.

Abu Talib al-Makki, a Sufi scholar of the tenth century, in his *Qut al-Qulub* (*The Nourishment of Hearts*), expresses the fundamental importance of hearing as the source of receiving the first knowledge of God. He declares that, in fact, "the only knowledge worth seeking is knowledge of God based on hearing rather than seeing."[10] Al-Ghazali, as well, in his influential book *Music and Singing*, treats this as a fundamental truth through which he explores the intimate connection between hearing of and longing for God and the relationship between sound in general and music in particular. He defines "the ear [as] the antechamber of the heart."[11] He continues:

[The] Beautiful voices and arrangements of notes [that mystics] hear during remembrance rituals as spiritual analogies are called music. The human's nature prefers these spiritual analogies to everything else. So when a person hears the analogies that pertain to notes, he tastes the truth, the original voice [of God], the sound of Unity. The taste of truth is received by every part of the body, the sight, the mouth, the intellect, arms and legs, and finally the heart, a destination that could not be reached by a thousand efforts other than hearing.[12]

Similarly, Ibn al-Arabi, one of the most influential mystical philosophers of the twelfth century, expresses that "the first we know of God, what was sent to us from Him [as knowledge], was his speech and pure listening." He adds,

"When the singer sings, ones worthy of hearing find, in their hearts, an un-veiling and intense yearning for that memory."[13] This last point is critical to the Sufi understanding of the meaning of music. Here, Ibn al-Arabi clearly establishes an organic connection between music and longing in the form of a call and response. The beautiful sound, called music, reminds human beings of their original love for God and fuels their yearning for Him. Thus the ob-jective of listening to music during separation is to arouse longing. Although longing is painful, in it there is a pleasure that lies in the hope of union. That is the meaning of pleasure to be drawn by listening to music. Al-Ghazali, in his *Alchemy of Happiness*, further expands on the meaning of music in rela-tion to longing, where he construes that "the best, most meaningful, and fun-damentally functional hearing of music is [when] . . . the listener's longing is aroused." A listening of this sort "strengthens of the listener's passion and love and . . . inflames of the tinderbox of his heart, and brings forth from it . . . states which he had not encountered before he listened to the music. . . . The cause of those states befalling the heart through listening to music is the secret of God."[14] Thus understood, music does not need lyrics to effect longing. It has a fundamental capacity to achieve yearning with its own "harmonies and beau-tiful properties."[15] At the same time, Sufi songs — *ilahis* in particular — are the most prevalent genre used in Sufi rituals. They are the musical settings of Sufi poetry, which, similarly, rests on the premise of endless longing for a beloved who can never be reached in this world. These songs at once express the yearn-ings of the poet and the composer and stimulate the yearning of the listener.

In addition, along with love, sorrow, and hope, a sine qua non faculty for longing is remembrance. Remembrance is one of the recurring themes in Sufi poetry because forgetting extinguishes the yearning. Finally, *ilahis* are repet-itive both in form and lyrics on the principle that repetitiveness is essentially related to remembrance. Repetition is sameness in linear time. It is being in the present, with a manifestation of having been there before. It is the upper-most expression of constancy and persistence in remembering God.

Unity of Dualities as Expressed in Sufi Poetry and Music

With this in mind, I shall now turn to an *ilahi*, a composition of a highly met-aphorical Turkish Sufi poem, to demonstrate how the pain of love and intense yearning for more love is expressed and exhibited. Yunus Emre, a thirteenth-century poet and Sufi mystic, writes a poem entitled "Dolap, niçin inlersin?" about a wheel whose only purpose and occupation is to turn above a well, to bring water from below up to the surface and pour it on the soil. The poem, made up of six four-line stanzas, opens with a single-line question directed to the wheel and its immediate answer:

Wheel, why do you sigh?
Of my sorrow, I sigh
I fell in love with my Lord
That is why I sigh

The rest of the poem, sixteen lines in total, extends and elaborates the answer of the wheel. We understand that the wheel considers its function to consist in remembrance and that it is grateful to God for destining it to remember Him in repetition and circularity, along with a deep sadness and pain for its ever-increasing love and longing for Him. At the last stanza Yunus, the poet, enters and claims ownership of the sorrow, since by this time we are sure that the poet has been expressing his own sorrow through the wheel metaphor.

Yunus-in-love says Ah!
The tears wipe away the sins
I swear to my love for God
That is why, I sigh

Yunus's poem characteristically expresses one central idea: love and yearning in submission and sorrow. Through the eloquently repeated utterances such as "sigh," "weep," and "love," Yunus's sorrow transmits itself and becomes the feeling of all his hearers. It is no longer his individual feeling; it is a collective sorrow and longing sparked by a collective remembrance.

Albay Selahaddin Efendi, the composer who set Yunus's poem to music, employs a repetition that encompasses both the simplicity and the repetitive circularity of the wheel in such a way that the song, musically as well, becomes a portrait for these concepts. In its overall structure the entire poem is set to the same melodic phrase introduced with the initial question and answer during the first four measures of the *ilahi*. On a smaller scale, the melodic phrase itself is internally repetitive and minimalistic. First, of the total twenty-four notes of the phrase, there are only four distinct pitches — D, C, B, and E. Also, its second and third measures are identical. Furthermore, the only difference between the first and the fourth measures is their final notes. It comes, therefore, to hearing the same, simple sound twenty-four times, over and over again, throughout the entire song. In addition, stanzas are separated by chorus sections added by the composer, offering traditional greetings to Prophet Muhammed. Each chorus section employs the initial melodic unit twice. By the time the song is completed the first four measures of the song are heard as many as thirty-six times. (Figure 1 shows the first of three stanzas and the refrain.)

Similarly, simplicity in repetition is an aesthetic and structural component of Pärt's music as well. In praise of simplicity, the composer once said: "All important things in life are simple."[16] After stating the pure beauty of the economy of expression as a compositional process, he claimed that he had discovered that "it is enough [even] when a single note is beautifully played." Pärt said that had he had the opportunity, he would have wanted to recapitulate the theory of tintinnabuli one day: "This theory has the same clarity as the structure of breathing: simple and tangible."[17]

Yunus Emre's poem connects to the fundamental theme of this paper, duality in unity, though covertly, at a deeper level. The wheel, a powerful metaphor for the helplessness of wistful waiting to return to God, is not alone in its remembrance and love. A mule (a wheel horse), which is tied to the wheel and circles around the well blindfolded, is the force that turns the wheel. The wheel without the mule is useless; a mule without the wheel is not necessary. Their destiny of endless revolving is locked together, and their purpose of bringing the water, the truth, to the surface is united.

Here, in essence, I find a fundamental convergence between the expression of longing in Sufi poetry and Pärt's tintinnabuli structure: Like the mule and the wheel, the voices remain separate in action yet one in purpose and longing. In my discussion of tintinnabuli, it was observed that seemingly binary oppositions — theological as well as musical — underlie the whole tension in the composition. The composer convincingly conveys the symbolic meaning of individual voices and their coexistence in two stages: First, he acknowledges that each voice has an independent meaning and function that seems to be in opposition. This constitutes the 1+1 part of the equation. There have been numerous metaphorical attributes in pairs that have been applied to Pärt's music and nested through tintinnabuli structure, such as suffering and consolation, straying and stability, sin and forgiveness.[18] While within these opposites one is always vulnerable and imperfect and the other is reliable and graceful, they all spring from one fundamental duality: human and divine. The M-voice is the human and the T-voice the divine, God. Pärt firmly clarifies the lower voice's identity: "The second voice is saying here 'without me, you can do nothing, nothing.'"[19]

After establishing the illusory duality of the human and the divine and the way human beings separate themselves from God by sinning, straying, and suffering, Pärt next asserts that "the two voices are in reality one voice, a twofold single entity."[20] This is the =1 of the equation. They are actually never

Do-lap ni - çin___ i - ni - ler - sin?___ Der - dim___ var - dır___ i - ni - le - rim.

Ben mev - la - ya___ â - şık___ ol - dum.___ A - nın___ i - çün___ in - i - le - rim.

Sub-han Al - lah,___ Sul - tan___ Al - lah,___ ya Nu - red - din___ Şey - hen - lil - Lah.

Han-nan Al - lah,___ Men - nan___ Al - lah,___ Her dert - le - re___ Der - man___ Al - lah.

Figure 1. Yunus Emre/Albay Selahaddin Efendi, "*Dolap niçin inilersin?*" First of three stanzas and refrain.

separated, because stability, forgiveness, and consolation hold firmly onto the straying, sinning, and suffering one. The voices' interaction unfolds around tension (straying) and resolution (stability), an oscillation that sustains the everlasting yearning to return to the initially experienced union. It is the manifestation of 1=1+1=1.

Mevlânâ Muhammed Celâleddin-i Rûmi, the fourteenth-century Sufi poet, powerfully depicts the concept of duality in union through a moth's annihilation in the flame of a candle. In his *Fihi ma Fihi*, he defines this relationship as existential, a form of call and response that is rooted in their nature:

> Whenever the moth casts itself into the candle, it burns — yet the true moth is such that it could not do without the light of the candle, as much as it may suffer from the pain of burning. If there were any animal like the moth that could do without the light of the candle and would not cast itself into this light, it would not be a real moth, and if the moth should cast itself into the candle's light and the candle did not burn it, that would not be a true candle.[21]

When I began to study Pärt's tintinnabuli, I expected to meet the Other. I found, instead, in the beauty of its structure and its unity of dualities and in his evocation of spiritual longing an illumination and a kinship. Pärt confirmed the existence of unity in sound beyond the musical structures where separation is no more.

My soul wearies for the Lord, and I seek Him in tears. How should I not seek Him? When I was with Him my soul was glad and at rest . . .

Adam's Lament[22]

Notes

1. Peter C. Bouteneff, *Arvo Pärt: Out of Silence* (Yonkers, NY: SVS, 2015), 28. Also see Robert Sholl, "Arvo Pärt and Spirituality," in *The Cambridge Companion to Arvo Pärt*, ed. Andrew Shenton (Cambridge: Cambridge University Press, 2012), 140–58; Andrew Shenton, "Arvo Pärt: In His Own Words," in *The Cambridge Companion to Arvo Pärt*, ed. Andrew Shenton (Cambridge: Cambridge University Press, 2012), 111–27; Brian Hehn, "Theological and Musical Techniques in Arvo Pärt's Tribute to Caesar," *Choral Journal* 57, no. 4 (November 2011): 32–42.

2. Paul Hillier, *Arvo Pärt* (Oxford: Oxford University Press, 1997), 96.

3. Arvo Pärt, conversation with Antony Pitts, BBC Radio 3, recorded at the Royal Academy of Music in London, March 29, 2000.

4. Marguerite Bostonia, "Bells as Inspiration for Tintinnabulation," in *The Cambridge Companion to Arvo Pärt*, ed. Andrew Shenton (Cambridge: Cambridge University Press, 2012), 138.

5. For more on the *bezm-elest* (primordial gathering) and comparative discussion based on the primary sources, see Suleyman Uludağ, *İslam Açısından Müzik ve Sema* (İstanbul: Kabalcı Yayınevi, 2005).

6. Abu Hamid al-Ghazali, *Love, Longing, Intimacy, and Contentment* (Cambridge: Islamic Texts Society, 2011), 88–98.

7. Al-Ghazali, *Love, Longing, Intimacy, and Contentment*, 88.

8. Al-Ghazali, *Love, Longing, Intimacy, and Contentment*, 89.

9. Al-Ghazali, *Love, Longing, Intimacy, and Contentment*, 90.

10. Abu Talib al-Makki, "The Nourishment of Hearts [Qut al-Qulub]," trans. Atil Khalil, in *The Muslim World* (Oxford: Blackwell, 2011), 18.

11. Abu Hamid al-Ghazali, "On Music and Singing," trans. Duncan B. MacDonald, *Journal of the Royal Asiatic Society of Great Britain and Ireland* (1902): 199. See also James Robson, ed. and trans., *Tracts on Listening to Music: Dhamm al-Malāhi by Ibn abī 'l-Dunyā, and Bawāriq al-ilmā' by Majd al-Dīn al-Ṭūsī al-Ghazālī* (London: Royal Asiatic Society, 1938).

12. Al-Ghazali, "On Music and Singing," 196.

13. William Chittick, *Ibn Al-Arabi's Metaphysics of Imaginations: The Sufi Path of Knowledge* (Albany: State University of New York Press, 1989), 289.

14. Abu Hamid al-Ghazali, *Alchemy of Happiness [Kimiya al-saadat]*, ed. Seyyid Huseyin Nasr, trans. Jay Crook (Chicago: Great Books of the Islamic World, 2007).

15. Al-Ghazali, *Alchemy of Happiness*, 89.

16. Geoffrey J. Smith, "Sources of Invention: An Interview with Arvo Pärt," *Musical Times* 140, no. 1868 (Fall 1999): 21.

17. Smith, "Sources of Invention," 22

18. Bouteneff, *Arvo Pärt*, 176–86.

19. A clear reference to the Gospel of John: "Apart from me you can do nothing" (15:5).

20. Hillier, *Arvo Pärt*, 96.

21. Mevlânâ Muhammed Celâleddin-i Rûmi, *Fihi ma fihi*, ed. Badi'ozzaman Foruzanfar, trans. A. J. Arberry (Tehran: Tehran University, 1961), end of discourse 10.

22. Text by St. Silouan the Athonite. See *Saint Silouan the Athonite* (Crestwood, NY: SVS, 1999), 448, set to music in Arvo Pärt's 2010 composition *Adam's Lament*.

Contributors

Andrew Albin is Associate Professor of English and Medieval Studies at Fordham University. His scholarship in the field of historical sound studies examines embodied listening practices, sound's meaningful contexts, and the lived aural experiences of historical hearers—in a word, the sonorous past—as an object of critical inquiry. His work has been recognized with grants and fellowships from the American Council of Learned Societies, the Medieval Academy of America, the Pontifical Institute of Mediaeval Studies, and the Yale Institute of Sacred Music. He is the author of *Richard Rolle's Melody of Love: A Study and Translation with Manuscript and Musical Contexts* (PIMS, 2018).

Peter C. Bouteneff is Professor of Systematic Theology at St. Vladimir's Orthodox Theological Seminary. He is co-founder and Director of the Arvo Pärt Project and Founding Director of the seminary's Institute of Sacred Arts. He has authored and/or edited a dozen books and some fifty scholarly essays in diverse fields. He has written extensively on Arvo Pärt in popular and academic essays and is the author of *Arvo Pärt: Out of Silence* (SVS Press, 2015). Since 2017 he has been on the Board of Artistic Advisors of the Arvo Pärt Centre, located in Laulasmaa, Estonia.

Maria Cizmic is Associate Professor in the Humanities and Cultural Studies Department at the University of South Florida, where she teaches courses that integrate musicology into an interdisciplinary curriculum. She is author of *Performing Pain: Music and Trauma in Eastern Europe* (Oxford, 2012). In addition to music and trauma studies during the late socialist period, her areas of research and teaching include twentieth-century American experimental and popular music; film music; disability studies; and performance, technology, and mediation. Her work can also be found in *Twentieth-Century Music, American Music,* and *Music, Sound, and the Moving Image*.

Jeffers Engelhardt is Associate Professor of Ethnomusicology at Amherst College, where he teaches courses focusing on community-based ethnography, music and religion, voice, and analytical approaches to music and sound. His research deals broadly with music, religion, European identity, and media. He is the author of *Singing the Right Way: Orthodox Christians and Secular Enchantment in Estonia* (Oxford, 2015) and coeditor of *Resounding Transcendence: Transitions in Music, Religion, and Ritual* (Oxford, 2016). His current book project is *Music and Religion* (under contract with Oxford University Press), and he is Editor-in-Chief of the *Yale Journal of Music and Religion*.

Adriana Helbig is an Associate Professor of Music at the University of Pittsburgh, where she teaches courses on global hip-hop, music and disability studies, and prison sounds. A recipient of fellowships from the American Council of Learned Societies, National Endowment for the Humanities, American Councils for International Education, IREX, and Fulbright, she has focused her research on issues relating to music and human rights. Her books include *Hip Hop Ukraine: Music, Race, and African Migration* (Indiana University Press, 2014), *Hip Hop at Europe's Edge: Music, Agency, and Social Change*, co-edited with Milosz Miszczynski (Indiana University Press, 2017), and *ReSounding Poverty: Romani Music and Development Aid* (forthcoming).

Paul Hillier's career has embraced singing, conducting, and writing about music. He was founding director of the Hilliard Ensemble and subsequently founded Theatre of Voices. He was Principal Conductor of the Estonian Philharmonic Chamber Choir (2001–2007) and has been Chief Conductor of Ars Nova Copenhagen since 2003. His recordings, consisting of over a hundred CDs, including seven solo recitals, have earned worldwide acclaim. Arvo Pärt has dedicated two of his compositions to Hillier, who in turn has made premiere recordings of several Pärt works. His books about Arvo Pärt and Steve Reich, together with numerous anthologies of choral music, are published by Oxford University Press. In 2006 he was awarded an OBE for services to choral music.

Kevin C. Karnes is Professor of Music and Vice Provost for the Arts at Emory University (Atlanta). His research explores sounding expressions of identity, difference, and belonging in Eastern and Central Europe from the nineteenth century to the present. His published work includes *Arvo Pärt's* Tabula Rasa (Oxford, 2017), *Jewish Folk Songs from the Baltics: Selections from the Melngailis Collection* (A-R Editions, 2014), and *Baltic Musics/Baltic Musicologies: The Landscape since 1991* (with Joachim Braun, Routledge, 2009). He is currently working on a new book, *Sounds Beyond: Arvo Pärt and the 1970s Soviet Underground*, and serving as Editor-in-Chief of the *Journal of the American Musicological Society*.

Alexander Lingas is a Professor of Music at City, University of London, founder and Musical Director of the vocal ensemble Cappella Romana, and a Fellow of the University of Oxford's European Humanities Research Centre. His present work embraces historical study, ethnography, and performance. In 2018, His All-Holiness, Bartholomew I, Archbishop of Constantinople–New Rome and Ecumenical Patriarch, bestowed on him the title of Archon Mousikodidaskalos.

Christopher J. May completed his doctorate in musicology at Oxford University in 2016. His dissertation examined a number of prominent critical frameworks for Pärt's music and sought especially to reconcile analytical and experiential accounts of tintinnabuli. In 2015 and 2016 he was the recipient of visiting studentships at the Estonian Academy of Music and Theatre. He holds a law degree from the University of Sydney, and his other research interests include music copyright law.

Ivan Moody, a world-renowned composer and conductor, is also a researcher at CESEM–Universidade Nova, Lisbon, and Founding Chairman of the *International Society for Orthodox Church Music*. He was Professor of Church Music at the University of Eastern Finland from 2012 to 2014. He is a contributor to the revised edition of *The New Grove Dictionary of Music*, *The Canterbury Dictionary of Hymnology*, *Musik in Geschichte und Gegenwart*, and the forthcoming *Cambridge Stravinsky Encyclopaedia*. He has published widely on music and spirituality and on music in the Balkans and the Mediterranean and is the author of *Modernism and Orthodox Spirituality in Contemporary Music* (ISOCM, 2014).

Robert Saler is Research Professor of Lutheran Studies at Christian Theological Seminary, in Indianapolis, where he also serves as Associate Dean and Executive Director of the Center for Pastoral Excellence. He is the author of several books and dozens of scholarly articles. His most recent book is on theology and Radiohead, and he is currently researching Christological themes in the music of David Lang.

Andrew Shenton is a scholar, prize-winning author, performer, educator, and administrator based in Boston, Massachusetts. He holds masters and doctoral degrees from Yale and Harvard, respectively, and is a Professor of Music at Boston University. His extensive work on Arvo Pärt includes *The Cambridge Companion to Arvo Pärt* (Cambridge, 2012), which he edited; the monograph *Arvo Pärt's Resonant Texts: Choral and Organ Music*, *1956–2015* (Cambridge, 2018); and several essays. He has conducted the US premiere of several of Pärt's works as Artistic Director of the Boston Choral Ensemble.

Toomas Siitan is an authoritative Pärt scholar, having worked closely with the composer for several decades. He graduated from the Estonian Academy of Music as a composer and received his PhD in musicology from the University of Lund. Since 1986, he has taught music history at the Estonian Academy of Music and Theatre, rising to Professor in 2004 and serving as head of the Department of Musicology since 2013. Siitan has published extensively on Pärt, Protestant singing and church history in nineteenth-century Estonia and Latvia, and the general history of Western music. Siitan's work has been recognized by the Estonian Composers' Union, the Estonian Cultural Foundation, the Estonian Music Society, and by a service medal from the Estonian government.

Sevin Huriye Yaraman, a music theorist, received her PhD from the Graduate Center of the City of New York. She is a Senior Lecturer at Fordham University, where she teaches Music for Dancers in the Fordham/Alvin Alley BFA program and seminars on music and dance and the perception of others. She was the recipient of an AAUW fellowship in support of her book *Revolving Embrace: The Waltz, as Sex, Steps, and Sound* (2002) and received a fellowship from the Istanbul Tarihi Turk Muzigi Toplulugu (Istanbul Historical Turkish Music Society) for her research on the role of music and movement in the Sufi contemplative practices of Turkey. Yaraman was granted a Faculty Fellowship for Fall 2020 to develop her current interest in Music and Disability.

Index of Terms

1+1=1: 7, 171, 225, 233–234, 238–239

acoustics: 4–5, 94, 103, 134, 136–137, 139–140, 144, 146–147
Adam (father of humankind): 16, 114
Amherst College: 7
apophasis: 109, 201, 207n11
Arvo Pärt Centre (APC): 7, 16, 41, 46, 60, 184, 227, 229n11
Arvo Pärt Project: 4, 6–7, 18
asceticism (ascetic practices): 110
Austria: 181

Baroque: 30
BBC Radio 3: 46, 65n22, 240 n3
Berlin: 202, 227
bezm-elest: 240n5
Bible, the: 203, 223
binaries: "coinciding opposites," 33; conformity/resistance, 52; consonance/dissonance, 190; corruption/morality, 44; healing/destruction (deconstruction), 200; human/divine, 238; idol/icon, 200, 204; intellectual/naïve, 33; life/death, 200; loud/soft, 9; melody/harmony, 30; movement/stasis, 30; official culture/counterculture, 44; oppression/resistance, 44; precious and delicate/strong and powerful: 89; presence/absence, 199–200, 202, 207n19; public self/private self, 44–45; simple (simplicity)/complex (complexity), 9, 180–181; sin/forgiveness: 238; sound (resonance)/silence,

30, 197, 199, 205; state/people, 44; straying/stability, 238–239; suffering/consolation, 18, 238; text/music, 30, 208; tonal/atonal, 9; tonic/dominant, 30; tradition/originality, 30
body, the (human): 4, 6, 8, 157, 177–8, 182, 185–186, 192, 213, 235
Bolshevik Revolution, sixtieth anniversary of: 76, 82
"bright sadness": 7, 234

canon, proportional: 27, 63
cantillation, scriptural: 15
Cappella Romana: 225
catharsis: 110
chant(ing): 32, 92, 117, 133, 158, 202, 221, 223–227, 233; Byzantine, 221, 224–226; Gregorian, 220–222, 225; liturgical, 7, 224, 227; plainchant (plainsong), 4, 15, 90–92, 103, 154, 221, 224–226, 230n41; Slavonic, 221, 224–225
chorale(s): 79, 92–93, 112
chords: 16, 90, 114, 118, 225; 6/3: 29; 8/5: 29
Christian Theological Seminary: 7
Christianity: 110, 205, 207n15; Orthodox, 19, 79, 108
Christology: 205
cinema: 46, 63, 177–178
collage: 25, 46, 220, 222
concert(s): 10, 30, 68–73, 75, 82, 83 n1, 103, 137, 186, 220
Cross, the: 79, 214

diatonicism: 29, 51
discotheque(s) (discos): 69, 71, 75–76, 78, 82
divine, the: 107, 109, 113, 158–159, 162–163,
    200, 202–203, 213, 238
dodecaphony (dodecaphonic music): 26, 33,
    46–47, 50–52
drone (ison; isokratema): 10, 114, 168–169,
    186, 221, 226, 230n40
dynamic(s): 15, 27–28, 60, 94–95, 99, 112, 114,
    124, 133, 163, 177, 191

ECM: 12, 46, 89, 99, 124, 136, 155, 181, 186,
    189, 192, 225
Eesti Telefilm: 37–41
Ephrata Community (Ephrata Cloister of the
    Seventh-Day Baptists): 32
epiklesis: 110
epoché: 133, 182
Eres Edition: 7
Estonia: 16, 37, 42, 135; Soviet: 28, 37, 42,
    66n49, 68, 100, 181
Estonian Composers' Festival: 30
Estonian Composers' Union (Union of
    Soviet Estonian Composers): 37, 43, 56,
    66n47, 70, 73
Estonia Concert Hall: 68
Estonian Film Archive: 38–41, 59, 61
Estonian Film Institute: 37
Estonian Philharmonic Chamber Choir:
    218n6, 244
Estonian Public Broadcasting: 37, 86n27
Estonian Theatre and Music Museum: 41–
    42, 48, 53, 57, 71, 83n1
ethnomusicology: 177
Europe: 5; Central and Eastern: 181; West-
    ern: 181

faith: 4, 6–7, 16–18, 31, 71, 75, 160–161, 169,
    197, 208; sound of, 19
fauxbourdon: 29
fermata: 121–122
film score(s): 5, 36–44, 47–48, 51–52, 56,
    60–63, 64n2; disavowal of, 43, 47; as "wage
    music," 36, 43
Fordham University Press: 7
formalism: 47

genre(s): 25–26, 30, 42, 44, 52, 115, 131, 179,
    226, 236,
Germany: 202; Nazi, 202; West, 181
Ghent: 227
glasnost: 81

Goeyvaerts Trio, the: 131, 148nn9,10
Gospel: of John, 33, 209, 241n19; of Matthew,
    211; of Mark, 211; of Luke, 211
"Great Religious Idea" (Velikaya Reli-
    gioznaya Ideya): 79

habitus: 178–179, 182, 187, 192
harmony: 26, 29–30, 32–33, 80, 95, 129, 158,
    161, 192, 206n8, 224–226, 230n39
Henry Luce Foundation: 7
hesychasm (hesychast tradition): 108–109,
    136, 154
hesychia: 108
Hilliard Ensemble (the Hilliards): 90, 99–
    100, 106, 128–129
homophony: 92
Hortus Musicus: 30, 68, 70, 72–73, 82, 83n1,
    230n27

ilahis: 236–237
inexpressible, the: 107, 124
Institute of Sacred Arts: 7
instrumentation: 12, 20n17, 31, 63, 170, 178–
    179, 185
Instrument(s): 12–14, 31, 56, 60, 76, 102, 114,
    119, 131, 133, 146, 153, 169, 177–179, 183,
    187, 192; bass, 52, 184, 206n8; bell(s), 94,
    105–106, 126n28, 135, 137–140, 151–152n52,
    154, 186; bells (Greek), 106; bells (Japa-
    nese) (rin), 106; bells (Russian) (zvon), 106;
    bells (tubular), 137–139, 151–152n52; bells
    (Western European), 106; cello(s): 28, 58,
    60–61, 67n52, 115, 166, 169, 186; chimes,
    52, 186; clarinet, 117, 149n15; crumhorn,
    230n27; drum, 52, 114, 206n8; guitar,
    151n51; harpsichord (cembalo), 56, 79;
    keyboard, electric, 63; organ(s), 91, 95, 114,
    122, 159, 194n35, 206n8; piano, 26–27, 33,
    41, 52, 58, 60, 67n52, 118, 129–131, 133, 140,
    141, 143, 144, 146–147, 147n2, 148nn5,6,7;
    151n51, 152nn58,60,61, 152–153n62, 177, 179,
    182–187, 189–191; piano (damper pedal),
    129, 148n15, 152n58, 184, 191–192; piano (sus-
    tain pedal), 182, 184–185, 192; saxophone,
    52, 63; strings, 79, 114, 119, 121, 129–130,
    140–141, 147–148n5, 184, 186; triangle, 114;
    trombone, 52; trumpet, 52; vibraphone, 52;
    viola, 115, 119; violin(s), 76, 115, 119, 186,
    230n27; violoncello, 115, 119
intent, authorial: 17–18
intonation: 93, 99, 136, 146; just, 131, 148n9
isokratema: 230n40. See also drone

kataphasis: 109, 201, 207n11
KGB: 81
kenosis: 110
key signature: 60, 223

Lancaster County, Pennsylvania: 32
Landini cadence: 29
Last Supper, the: 214
Latvian Centre for Contemporary Art: 71–73, 77, 81
Latvian SSR Philharmonic Chamber Orchestra: 75
Latvian State Library and Conservatory: 76
Laulasmaa, Estonia: 229nn11,27, 230 n40
logogenesis (logogenic): 11, 15, 226, 230n41
logos: 12, 14, 16, 33, 171
longing: 154, 160, 163, 232–238, 240
Los Angeles Philharmonic: 31
love: 114, 158–159, 184, 201, 209, 232, 234, 236–237

mass(es) (liturgies): 17, 56, 58, 70, 74–75
mathematical order (technique, pattern, logic, method): 27, 28, 33, 130
melody: 180, 186–187, 189–190, 221, 223–227
minimalism: 9–10, 100, 104, 172n6; "holy," 221; social, 180, 184
Ministry of Culture: 68–69, 78
modernism: 5, 181, 185
monophony: 220–221, 223–227
music: 6, 9, 11–19, 20n5, 28–34, 36–37, 41–46, 50, 52, 54, 56–57, 60, 62–63, 67n59, 69, 71, 74–76, 79–80, 82, 89–122, 124, 131, 133–137, 139, 146–147, 154–158, 160–171, 177–187, 190–193, 197–202, 205–206, 206n8, 208–209, 211, 213–214, 218, 220–227, 229n27, 232–238, 240; angelic, 158, 231n45; as translation, 17–19; avant-garde, 25, 28, 33, 66n49, 81; children's, 51, 65n32; Christian, 211; comprehensibility of, 15–16, 181–182; concert, 54–55, 58, 63; dance, 9; devotional, 70; early, 4, 26, 30, 51, 91–93, 100, 187, 191–192, 220–222; elements, horizontal and vertical, joined, 27; embodiment (sonic), 3, 116, 133, 139, 146, 177, 184–185, 193, 202; European, 29; expressivity, 181–182, 184; fifths, cycle of, 51; film, 5, 26, 36–37, 43–47, 47, 54, 56, 60, 62–63, 220; golden age of, 205; golden section, 27; medieval, 80, 90, 92, 154, 172, 220, 225; modal, 212, 223–224, 226; modernist, 100, 181; pandissonant, 30, 187; Renaissance, 90–92,

154, 157, 220, 222; sacred, 4, 6–7, 58, 66n49, 74, 79; seconds, dissonant, 30; seventh(s), 30, 130, 134; Soviet, 26, 44, 45, 76; film, 43, 70; spiritual, 157; theater, 26; timbre(s), 131, 133–134, 140, 146, 151n51, 182–183; tintinnabuli (*see below*); tonal, 105, 118, 179, 183, 191–192, 223–224, 226; as translation, 17–19; twelve-tone, 26–29, 45, 47–48, 50–53, 181; vocal, 29–31, 56, 90, 102, 178, 221, 225; Western concert, 29–30, 147
musicology: 3, 177, 205
mysticism: 173n13; Christian, 136, 157; musical, 157–162

narrative, archetypal: 46–47, 51
National Archives of Estonia: 59, 61
neoclassicism: 26
North America: 32
note length: 16
note placement: 11

oppression, Soviet: 33, 44
orchestra, string: 91, 169
Orthodox Christian Studies Center (Fordham University): 7
Oxford, England: 157, 173 (n11)

paradise: 16, 114, 217
"Pärt phenomenon," the: 4, 6, 199
"Pärt sound," the: 5, 8–12, 14, 19n3, 89
performance(s): 58, 68–69, 71, 74–75, 78–79, 91, 98–100, 102, 105, 108, 111, 114, 118–122, 124, 131, 134–135, 161, 165, 169, 174n22, 177–179, 184–187, 192, 194n27; Byzantine, 225; cultural ritual, 54; musical, 179, 183; space, 118
phenomenology: 3, 161, 177–178, 185
plane, harmonic: 29
poetry: Greek epic, 15; Sufi, 7, 236, 238
polyphony: 26, 90–94, 103, 180, 220, 223, 225–226; Flemish, 29–30
prayer(s): 31, 109, 134, 157–158, 169–170, 201–202, 204, 211, 223; hesychastic, 154; Jesus, the, 223, 227, 229n23; liturgical, 17; Orthodox (Christian), 74
propaganda, religious: 81
Psalm(s): 31, 117, 121, 158, 169, 180, 202, 223–224, 229nn19,26
Psalter, the: 224
psychoacoustics: 130–131, 133, 136, 140–141, 143, 146, 148n8, 149n17, 161
psychology: 4–7, 118

quartet, string: 31, 78, 120
Qur'an, the: 233–234

realism, socialist: 50, 66n36, 181, 192
reception: 6, 8, 16, 42, 46, 63, 83n2, 157,
    179–180, 192
recitatif: 15
religion: 71, 74, 78, 83n2, 108, 136, 147, 192,
    201, 223
resonance(s): 7, 12, 17, 29, 37, 118, 121, 130–131,
    134, 136, 139–141, 143, 146, 148n5, 153n62,
    179, 182, 184, 186, 189–190, 197–199, 209
rest: 103, 112, 116–118, 120–124, 168
rhythm: 11, 93, 108, 133, 182, 224, 227; articu-
    lation, 100; complexity, 28; patterns, 91–92;
    proportions, 27; regularity, 60; system of,
    32
Riga, Latvia: 66n49, 69, 70–71, 75, 78–81, 83
Riga Polytechnic Institute: 81–82; student
    club, 69, 71, 75–76, 78, 80–82
ritornelli: 31
"ring": 93–94, 96, 105
Rome: 227

sacred, the: 4 ,7, 19
St. Vladimir's Orthodox Theological Semi-
    nary: 4, 7
salvation: 109, 205, 207n15, 211–213
scale: 96, 105, 214; C-major, 28; diatonic, 26,
    30, 105; natural minor, 30; overtone, 129;
    triad, against, 62
scores: 52–53, 55, 59, 120–122, 124, 130, 163–
    165, 168–170, 182, 190; graphic elements of,
    20n11, 98, 166, 169, 189
serialism: 25–26, 29, 32, 50
Sermon on the Mount: 32
Sigur Rós (Icelandic band): 19n5
silence(s): 4, 30, 46, 93, 103–104, 107–124,
    131, 133–137, 140–141, 143, 146–147, 154,
    156–157, 159, 162–163, 168, 170, 179–180,
    184, 191, 197–199, 204–205, 208–209, 211,
    217–218, 220, 232; acoustics, 134; breathing,
    116; notation, 119–124; performance, 119–
    124; phrases, delineation of, 116; power,
    109; pre-tintinnabuli works, 111; structural,
    112–113, 116; surrounding, 113–115
simplicity, 11, 28–29, 33–34, 51, 100, 179–183,
    187, 191–192, 206, 233, 237–238
singing, liturgical: 221
Song of Symeon: 212
sonic imaginations: 198
sonic landscape: 9

sonic world(s): 10, 12, 37, 177
sound: 8, 12–13, 30, 33–34, 197–198; as me-
    dium, 9, 12; decay of, 130–131, 139, 143,
    148n7, 179, 182, 184–186, 188, 191–193; pre-
    condition of, 8
"Sounding the Sacred" (conference): 4
soundscape(s): 25, 30, 60, 63
soundtrack(s): 9, 15, 26, 42, 57, 63, 70, 78,
    105, 182
Soviet Latvian Composers' Union: 81
Soviet Union: 36–37, 44, 51, 54, 71, 74
Soviet Youth League (Komsomol): 76, 78, 81
spirituality: 110, 136, 147, 157–158, 161, 180
stasis: 30, 204, 226
Stockholm: 227
studies: cultural, 4, 6; materiality, 4; media,
    3, 6; music, 3, 4, 6; Pärt, 4, 15, 46, 145, 197–
    198, 201, 205; performance, 4; religious, 6;
    sound, 3–6, 133, 149n17) 175n26, 197–199
Sufism: 232–234

taksim: 224
Tallinn: 30, 42, 68–69, 71, 75, 80–82, 83n3,
    84nn19,27), 85n54, 131, 186, 220
Tallinn Chamber Choir: 68
Tallinn Conservatory: 36
Tallinnfilm: 37–39
Tartu: 30, 69–70, 8n3
Tartu, University of: 25, 68–69
tessitura: 112
text(s): Estonian, 223, 227; Greek, 227; Latin,
    32, 73–75, 157, 164, 215, 217–218, 223; Rus-
    sian, 16, 31, 223; sacred, 6, 16–18, 21n25,
    58, 66n49, 75, 170, 221; silent, 16, 31–32,
    126n25; Slavonic, 15–16, 31, 169, 221, 223–
    225, 227
Thaw, the: 45
themes: consolation, 18, 199, 238; death, 18,
    200; despair, 19, 117, 119; hope, 18, 117, 119;
    life, 18, 200; loss, 18, 43; suffering, 19, 108,
    110, 211, 214, 224, 238
theology: 4, 6–7, 108, 110, 112, 197–199,
    201–206, 208, 233–234; Christian, 4, 6, 147,
    199–200, 208; Orthodox, 136, 174n22; Sufi
    Islam, 4, 232–235
theosis: 110
time signature: 60, 90
tintinnabuli (tintinnabulation): 3–7, 10–11,
    13, 20n7, 25–26, 28–30, 33, 36–37, 42, 46–
    47, 51–52, 54–58, 60–63, 66nn49,50, 68–71,
    73–75, 78–79, 82, 83n17, 89–93, 96–99, 101,
    105–106, 109, 111–112, 115–118, 121–122, 129,

133–137, 139–141, 143, 146–147, 154, 170, 172n6, 176n32, 177, 179–182, 185–187, 190, 192–193, 205–206, 208, 217, 220–227, 228n6, 230n40, 232–233, 238–239; acoustics of, 5, 8, 129; algorithm(s) of, 74, 133, 137, 140, 146; barring in, 90–91; central pitch in, 60, 66n50, 116; compound, 60, 62; dissonance in, diatonic, 180, 190, 192; expression marks for, 96–98; flexibility, 98; M-voice (melodic voice), 58, 60, 62, 66n50, 67n52, 92–94, 96, 101, 105, 168, 170, 186, 220–221, 225, 233, 238; pitch, primacy of, 95; principles, 9, 28, 32, 58, 62, 66n50, 74, 116, 134; "prototintinnabuli," 28; silence in, 103, 107–108, 131, 137, 139–140; T-voice(s) (triadic voice), 58, 60, 62, 66n50, 67n52, 74, 92–93, 96, 101, 105, 168, 170, 186, 220, 225–226, 230n40, 233, 238; triad(s), 9–10, 13, 26, 28, 62, 66n50, 90, 93–94, 96, 101, 105, 117, 134–135, 137, 140, 168, 180, 188, 220, 233
Tolleshunt Knights (UK): 227
tonality: 10, 104, 135, 179–181, 187, 192, 220, 226; abstract, 104
topos: 10–11

triadic harmony: 26, 29
Trinity, the (Holy): 109, 160, 208, 213, 214

unio mystica: 34
United States: 5, 181
Universal Edition: 7, 96, 164–166, 169, 185
University of Berlin: 202
University of Oregon: 29

Vatican: 20 (n13)
Venice: 27
via negativa: 110
vibration(s): 6, 34, 131, 133–134, 136–137, 140–141, 143–144, 146–147, 150n27, 151n52, 184, 206, 235
vibrato: 94–95, 100, 131, 134, 146
Vilnius String Quartet: 78
vox humana: 95

wave(s), sound: 3, 8, 157, 161
Warsaw Autumn Festival of Contemporary Music: 28

Zurich: 227

Index of Persons

Abo-Saif, Boshra: 41
Abraham: 214
Adorno, Theodor: 33, 180
al-Ghazali, Abu Hamid: 234, 236
al-Makki, Abu Talib: 235
Albin, Andrew: 4–5, 172n8, 173n15, 174n22
Artuoja, Virve: 38, 40
Artusi, Giovanni: 80
Athanasius, St.: 223
Atteln, Günter: 20n13, 184
Avramecs, Boriss: 76, 80, 82, 85n47

Bach, Johann Sebastian: 17, 26, 92–93, 222
Barth, Karl: 203–204
Begbie, Jeremy: 180, 205
Behr, John: 211–212
Beissel, Conrad: 32–33
Berger, Karol: 226
Bērziņa, Aina: 75, 81–82
Björk: 12
Blackwell, Albert: 206 (n8)
Boccherini, Luigi: 182
Boethius: 176n37
Boiko, Martin: 82
Bonhoeffer, Dietrich: 202–205
Bouteneff, Peter: 5, 7, 25, 64n9), 89–106,
    108–109, 125n7, 136, 147, 153n70, 172n5,
    174n22, 175n28, 176n34, 199, 205, 206n6,
    207n11, 228n10, 232
Bowers Broadbent, Christopher: 95
Brauneiss, Leopold: 10, 26. 30, 46, 64n9, 115,
    176n32, 217, 225

Buck, Nick: 207n8
Byrd, William: 92–93

Cacciari, Massimo: 213–217
Cage, John: 3, 104
Cerquiglini, Bernard: 166, 175n3
Chion, Michel: 133, 149n16
Cizmic, Maria: 5, 131, 179, 187, 226
Clements, Andrew: 211
Cleobury, Stephen: 122
Cobussen, Marcel: 108

Davydova, Lydia: 71
Davis, Adam: 198, 206n2
Denisov, Edison: 44
Derrida, Jacques: 174n25
Desmond, William: 201, 206
Dolan, Emily: 183
Dolp, Laura: 114, 136, 180, 227n1
Du Fay, Guillaume: 29–30

Egorova, Tatiana: 44
Eicher, Manfred: 12–13, 186, 192
Emre, Yunus: 236, 238
Engelhardt, Jeffers: 5, 20n5
Ephrem the Syrian, St.: 212

Fourier, Joseph: 130

Garbarek, Jan: 63
Gellrich, Jesse: 175n25
Gillespie, Vincent: 168

Glass, Philip: 211
God: 16–19, 33, 71, 74, 78, 80, 108–114, 157–
    160, 162, 169, 197, 199–200, 202–204, 206,
    207n15, 208–209, 211, 213–214, 217, 232–238;
    absence of, 208; union with, 110, 157
Gregory of Naxianzus, St.: 217–218
Grindenko, Tatiana: 84n23
Guardini, Romano: 209
Gubaidulina, Sofia: 33, 44

Hakobian, Levon: 44–45, 51–52
Haydn, Joseph: 183
Heidegger, Martin: 183, 191
Helbig, Adriana: 5, 108, 131
Helmholtz, Hermann von: 130, 148n5,
    152nn58,62
Hillier, Paul: 5, 27, 29, 36, 43, 45–46, 64n9,
    66n50, 89–106, 108–109, 118, 122, 136,
    139, 150n35, 151n48, 184–185, 214, 220–221,
    228n1
Hilton, Walter: 173n13
Himbek, Valdur: 41
Holy Spirit: 110
Hunt, Kadri: 51
Huxley, Aldous: 107

Ibn al-Arabi: 235–236
Ihde, Don: 182, 190–191

James, David: 129
Järvi, Neeme: 10, 14
Jesus Christ: 16–18, 32–33, 74, 79, 95, 108,
    110, 155, 169, 203, 205, 207n15, 208–209,
    211–214, 218, 223, 227, 229n23; Word (of
    God), the: 16, 33, 110, 169, 203, 208–209,
    212–213; Creator of the Universe: 209; Cru-
    cifixion: 211, 214; Incarnation: 208–209, 211,
    217–218; Passion: 211–214; Resurrection:
    211–212, 214–215, 217
Jõgi, Ants: 58
Josquin des Prez: 154, 222

Kabakov, Ilya: 69, 84n23
Kähler, Andreas Per: 108
Kaljuste, Tõnu: 124
Karnes, Kevin: 4, 37, 47, 58, 66nn46,49, 136,
    194n27, 231n45
Kask, Mati: 39–40
Katkus, Donatus: 76, 78
Khodorkovsky, Mikhail: 31
Kim-Cohen, Seth: 198, 204–205

Klas, Eri: 135
Khrennikov, Tikhon: 50
King, Martin Luther, Jr., Dr.: 31
Kivirähk, Ants: 40
Koppel, Virve: 41
Kõrver, Kristina: 48, 65n27, 229n11
Kromanov, Grigori: 40

Husserl, Edmund: 133, 182, 188

Laius, Leida: 26, 37–40, 58, 65n3
Lediņš, Hardijs: 69, 71–73, 75–76, 78–82
Le Guin, Elisabeth: 182
Lem, Stanislaw: 56
Lilje, Maia: 135
Lingas, Alexander: 7, 231n44
Lubimov, Alexei: 69, 71, 78, 81–82
Luther, Martin, 203

McClowry, Sean: 230n33
MacKendrick, Karmen: 197, 199
Macleod, Donald: 46
MacMillan, James: 180
Maimets-Volt, Kaire: 19n4, 37, 63, 64n5
Maler, Anabel: 60
Mamers, Jaak: 41
Mann, Thomas: 32
Manzoni, Giacomo: 27–28
Maran, Rein: 39–40
Marion, Jean-Luc: 200
Martynov, Vladimir: 73, 79–82, 85n45
Machaut, Guillaume de: 90, 154, 222
Maskew, Doug: 64n6, 84n19
Maximos the Confessor, St.: 209
May, Christopher J.: 4, 26, 70
Mellers, Wilfred: 213, 217
Methley, Richard: 173n13
Mihkelson, Immo: 65n24, 68, 135
Monteverdi, Claudio: 80
Moody, Ivan: 7
Moses: 200
Mowitt, John: 149n17
Muhammed, the Prophet: 237
Mustonen, Andres: 68, 73, 82, 83n1, 85n54,
    135, 230n27

Nelson, Dale: 211
Neuland, Olav: 39, 41

Nicholas of Cusa: 33
Nono, Luigi: 52
Normet, Leo: 108
Norton, John: 173n13

Obrecht, Jacob: 222
Ockeghem, Johannes: 17, 90, 222

Paistik, Avo: 40
Palestrina, Giovanni Pierluigi da: 17, 222
Pars, Heinu: 38–40
Pärt, Arvo: *passim*; aesthetic(s) of, 33, 46,
    51–52, 58, 70, 79, 93, 95, 106, 108, 111,
    113–136; childhood, 13, 184; children,
    music for, 15, 25–27, 33, 42, 51–52, 65n3;
    Christian, Orthodox, 17, 19, 108–109,
    135–136; composer, Christian, 70; com-
    poser, world's most-performed living, 3;
    controversy, twelve-tone, 36; "Creative
    laboratory," 26, 47–48, 52, 55–56, 58,
    62–63; *credo* crisis (scandal), 36, 74, 78,
    82, 84n25; diaries, 13, 58, 66n46, 67n51,
    69, 74–75, 78, 84n26, 135, 139, 180; em-
    igration to the West, 25, 36, 60, 62–63;
    engineer, recording (sound), 11, 192;
    ensembles, preferred, 13; faith, 6–7, 16,
    18–19, 31, 75, 208; genre, use of as tool of
    commentary, 52; humor, 99; "monk," 17,
    155, 172n5; "mystic," 5, 17, 154–155, 165;
    physicality, 182; piano, relationship with,
    177, 179, 182; record label (producer), 13,
    46; rehearsals, involvement in, 98; seri-
    alism, 25, 26, 30, 43, 50; silence, 4, 30,
    103–104, 107–109, 110–112, 133, 135–136,
    162, 180; "silent" period, 5, 25, 46, 54, 92,
    136, 154, 179, 187, 220–221; statements,
    political, 31–32; theater, composer for,
    19n4, 25–27; thinker, Christian, 113;
    training, 13; soundworld, 63, 69, 108,
    124; syllabic method (rules, shapings,
    principles), 15–16, 74, 58, 84, 90; "tintin-
    nabuli revelation," 31, 36, 220; transition
    years, 25, 29
Pärt, Nora: 56, 70, 73, 83n11, 99, 135–137,
    139, 155, 185, 187, 221–225, 229n11, 230n40,
    233
Phillips, Peter: 221
Piestrak, Marek: 40
Põldre, Mati: 41
Politkovskaya, Anna: 31
Prokofiev, Sergei: 44

Quinn, Peter: 51

Radakovich Holsberg, Lisa: 7
Rahaim, Matthew: 178
Randalu, Ivalu: 180
Rannap, Rein: 135
Reeves, Nicholas: 7
Restagno, Enzo: 28, 44, 118, 179–181, 190
Roeder, John: 225
Rolle, Richard: 156–163, 170–171, 173nn11,13,
    175n18
Roospipuu, Hans, 39
Rublev, St. Andrey: 213–214
Rûmi, Mevlânâ Muhammed Celâleddin-i:
    232, 239
Rustaveli, Shota: 25

Sabbe, Hermann: 185
Sakharov, Archimandrite Sophrony: 31,
    206n6
Saler, Robert C.: 7, 153
Sandner, Wolfgang: 46, 135, 154, 180
Savall, Jordi: 222
Schaeffer, Pierre: 133, 149n15
Schiller, Friedrich: 33
Schmelz, Peter: 43, 45, 50–51, 68, 181
Schnittke, Alfred: 44
Schoenberg, Arnold: 30
Scholl, Robert: 172n6, 185, 205–206
Schütz, Heinrich: 30
Schvak, Toomas: 289n11
Selahaddin Efendi, Albay: 237
Seraphim of Sarov: 223
Shenton, Andrew: 5, 7, 43, 131, 134, 136,
    205–206
Shostakovich, Dmitri: 44
Siitan, Toomas: 4, 68, 126n25, 135–136
Sildre, Joonas: 7
Silouan the Athonite, St.: 31, 112, 206n6,
    241n22
Silvestrov, Valentin: 78
Skowronec, Tilman: 183, 186
Small, Christopher: 137
Sontag, Susan: 110–111
Sööt, Andres: 39–40
Steinway, Theodore: 148n5
Sterne, Jonathan: 19n1, 146, 198
Sudnow, David: 179
Supin, Dorian: 139, 182, 227
Sviridov, Georgy: 44
Symeon, St.: 212

Taverner, John: 205
Thomson, William: 225
Tooming, Jaan: 40
Toop, David: 204
Tuganov, Elbert: 38–40
Tüür, Erkki-Sven: 135

Vaitmaa, Merike: 75, 135
Valk, Heinz: 39
Verdery, Katherine: 45, 65n20
Visocka, Asja: 76, 82

Volkonsky, Andrei Mikhaylovich: 44
Vorslavs, Lauris: 84n18
Vuorinen, Emppu: 108

Wagner, Richard: 90
Wilson, Robert: 184

Yaraman, Sevin Huriye: 7
Yurchak, Alexei: 44–45, 51–52, 54, 74

Zumthor, Paul: 166, 175 (n31)

Works by Other Composers

Cage, John, 4'33": 3

Du Fay, Guillaume Du Fay, *Missa L'homme armé*: 30

Martynov, Vladimir, *Passionslieder*: 79–82

Tchaikovsky, Pyotr Ilyich, *Sweet Daydream*: 33

Works by Arvo Pärt

Aatomik (Atom-Boy) soundtrack: 39
Aatomik ja jõmmid (Atom-Boy and the Thugs) soundtrack: 39
Adam's Lament: 16–17, 112, 114, 240, 241n22
Aetos: 67n59
Alleluia-Tropus: 115
An den Wassern zu Babel saßen wir und weinten: 75, 96, 99, 120
And One of the Pharisees: 121, 123
Annum per annum: 60–61, 67n53, 114
Arbos: 12, 69–70, 73
Arinushka variations (Variations for the Healing of Arinushka): 60, 62, 70

Beatitudes, The: 15
Beatus Petronius: 122–123
Berliner Messe: 31, 91, 115, 215, 217
Briljandid proletariaadi diktatuurile (Diamonds for the Dictatorship of the Proletariat) soundtrack: 40

Calix: 66n45, 75, 83n3, 84n26
Cantate Domino canticum novum: 31, 56, 58, 69–70, 73, 75, 99, 113
Cantus (in Memory of Benjamin Britten): 9–10, 62–63, 70, 115, 126n28, 137, 139, 182, 190
Cecilia vergine romana: 115
Collage über B-A-C-H: 26, 55
Como cierva sedienta: 114
Credo: 25, 32, 36, 45–46, 51, 74–75, 78, 84n25, 220

Da pacem (Domine): 12, 29, 92, 228n6
Dance of the Ducklings (Pardipoegade tants): 26–27
De profundis: 31, 70, 99
Deer's Cry, The: 9, 15, 115, 120
Diagramme: 26
Dopo la vittoria: 9

Enderby valge maa (Enderby-White Land) soundtrack: 39
Es sang vor langen Jahren: 114–115
Estländler: 115
Evald Okas (Evald Okas) soundtrack: 38

Festina Lente: 115
Four Easy Dances for Piano: Music for Children's Theater (Neli lihtsat tantsu klaverile. Muusika lastenäidenditele): 26–27
Fratres: 9, 12, 20n17, 62, 69–70, 73, 105, 185
Für Alina (Aliinale): 9, 11, 20n11, 46, 65 n25, 70, 83n3, 129–133, 139–143, 145, 148n8, 177, 179, 180, 185–192
Für Lennart in memoriam: 113

Habitare fratres in unum: 115
Helin (Ringing) soundtrack: 39
Hiirejaht (Mouse-Hunt) soundtrack: 38

In principio: 9, 33, 209
In spe: 73, 75, 79, 83 (n3), 96
Inimesed arsenalis (People in the Arsenal) soundtrack: 41

257

Inimeselt inimesele (From Person to Person)
    soundtrack: 39
Inspiratsioon (Inspiration) soundtrack: 41
Italian Concerto: 47

Jääriik (The Ice Realm) soundtrack: 39
Jäljed lumel (Footsteps in the Snow)
    soundtrack: 40, 58–62, 67n51
Just nii! (Just So!) soundtrack: 38, 52–55

Kanon pokajanen: 15, 115
Kapten (Captain) soundtrack: 41
Kaugsõit (The Long Journey) soundtrack: 40
Kind of Hush, A soundtrack: 62–63
Kolm portreed (Three Portraits) soundtrack: 41
Kurepoeg (The Young Stork) soundtrack: 39

Lamentate: 31, 114, 118, 131
Laul armastatule (Song to the Loved One): 25
Litany: 63, 114
Littlemore Tractus: 29, 114
Looduse hääled (Voices of Nature)
    soundtrack: 40

Maailma samm (Stride of the World): 45, 50
*Mäeküla piimamees (The Milkman of the
    Manor)* soundtrack: 38
Magnificat: 10, 105, 122
Meie aed (Our Garden): 51
Mein Weg (hat Gipfel): 63, 114
Mis? Kes? Kus? (What? Who? Where?)
    soundtrack: 39
Miserere: 62–63, 103, 116–117, 119
Missa syllabica: 11, 26, 31, 56–57, 74–75, 78,
    84n20, 111–112, 117, 120–122, 124, 226
Modus: 65n25, 69, 71, 75, 83n3, 84n13, 96–97

*Navigaator Pirx (Pilot Pirx's Inquest or, The
    Test of Pilot Pirx)* soundtrack: 40, 56–58,
    60, 66n49, 70, 78
Nekrolog: 36, 45, 50, 52
Nunc dimittis: 115, 131, 212

"O Adonai": 168–169
"O Sproß aus Isais Wurzel": 165–166
"O Weisheit": 165, 168
Õed (Sisters) soundtrack: 40
*Õhtust hommikuni (From Evening to Morn-
    ing)* soundtrack: 38
*Operaator Kõps kiviriigis (Cameraman Kõps
    in the Land of Stones)* soundtrack: 39
*Operaator Kõps marjametsas (Cameraman
    Kõps in the Berry Forest)* soundtrack: 38

*Operaator Kõps seeneriigis (Cameraman Kõps
    in the Land of Mushrooms)* soundtrack: 38
*Operaator Kõps üksikul saarel (Cameraman
    Kõps on a Desolate Island)* soundtrack: 38
Orient & Occident: 9–10, 115
Otsin luiteid (Searching for Dunes)
    soundtrack: 39

Pallid (Balls) soundtrack: 39
Pari intervallo: 12, 70, 73, 83n3
Partita: 115
Passio: 11, 31, 91, 95–96, 101, 105, 116–117, 120,
    205, 213–215, 217, 226
*Passio Domini nostri Jesu Christi secundum
    Joannem:* 31
Peace soundtrack: 41
Perpetuum mobile: 27–28, 48, 50, 52, 63
Pro et contra: 28
Psalom: 31, 63, 112, 121–122, 169
*Putukate suvemängud (The Insects' Summer
    Games)* soundtrack: 39

Quintettino: 26, 115

Rachel River soundtrack: 62–63, 67nn56,60
Reekviem (Requiem) soundtrack: 41
*Reportaaž telefoniraamatu järgi (Journalism
    by the Telephone Book)* soundtrack: 41

Saara: 71, 75, 84n16
Salve Regina: 115
Sarah Was Ninety Years Old: 9, 47, 50,
    65n25, 69, 96–97, 125n5
Scala cromatica: 115
Sellised lood (Such Stories) soundtrack: 40
Sequentia: 11
*Sieben Magnificat-Antiphonen (O-Anti-
    phonen):* 115, 164–168
"Silentium": 9–10, 115, 135
Silouan's Song: 31, 112–113, 115–116
Solfeggio: 9, 28, 30, 51, 104–105
Spiegel (im Spiegel): 9–10, 19n4, 20n21, 60–
    61, 67n51, 105
Stabat Mater: 11, 31, 115, 118–119, 121, 131, 214
Summa: 11, 56, 58, 69–70, 73, 75, 215, 217
Symphony No. 1 *(Polyphonic):* 26–27, 48
Symphony No. 2: 33, 115
Symphony No. 3: 25, 29, 92, 115, 220, 222
Symphony No. 4 *("Los Angeles"):* 31, 115

Tabula rasa: 9, 12, 20n7, 28, 62, 69–70, 89,
    115, 135–136, 180, 185
Täheke (Starlet) soundtrack: 40

Te Deum: 31, 62, 114, 116–117, 124, 226, 230n40

Test: 40, 56. 58, 66n49, 69–71, 73–74, 78–79

Tintinnabulum 1: 71, 84n13

Triodion: 115

Trisagion: 31, 169–170

Trivium: 70, 83n3

Ukuaru (Ukuaru) soundtrack: 39

Ukurtu valss: 37

Untitled for Saxophone and Keyboard: 67n59

Väike motoroller (The Little Motor Scooter) soundtrack: 26, 38, 48, 50–52, 54, 56, 62

Värvilised unenäod (Colorful Dreams): 40, 47, 50, 56

Värvipliiatsid (Colored Pencils) soundtrack: 40

Veealused (Sea-creatures) soundtrack: 40

Viimne korstnapühkija (The Last Chimney-Sweep) soundtrack: 38

Wallfahrtslied, Ein: 10, 115

Wenn Bach Bienen gezüchtet hätte (If Bach Had Kept Bees): 26, 56, 60, 71, 83n3

Woman with the Alabaster Box, The: 47, 115

Zwei beter: 115